COLLINS COUNTRYSIDE SERIES
PLANT LIFE

In the same series

LIFE ON THE SEA SHORE *John Barrett*

BIRDS *Christopher Perrins*

WOODLANDS *William Condry*

ROCKS *David Dineley*

INSECT LIFE *Michael Tweedie*

These books are intended to offer the beginner a modern introduction to British natural history. Written by experienced field workers who are also successful teachers, they assume no previous training and are carefully illustrated. It is hoped that they will help to spread understanding and love of our wild plant and animal life, and the desire to conserve it for the future.

COLLINS COUNTRYSIDE SERIES

PLANT LIFE

*

C. T. PRIME

with line drawings by
MARJORIE BLAMEY

and diagrams by
DIEDRE YUILL

READERS UNION
Group of Book Clubs
Newton Abbot 1978

CONTENTS

LIST OF ILLUSTRATIONS

between pages 64 and 65

7

12. Bell heather, heather (*Peter Wakeley*)
    Cross-leaved heath (*John Markham / J. Allan Cash*)

13. Yellow deadnettle, black mullein, eyebright (*Peter Wakeley*)

14. Foxglove (*John Markham / J. Allan Cash*)

15. Red deadnettle (*John Markham / J. Allan Cash*)
    Wild basil, bugle, cut-leaved germander (*Peter Wakeley*)

16. Clustered bellflower, nettle-leaved bellflower, harebell (*Peter Wakeley*)

17. White bryony (*J. Allan Cash*)
    Yellow bedstraw, woodruff (*Peter Wakeley*)

18. Teasel, scabious, perfoliate yellow-wort, felwort (*Peter Wakeley*)

19. Ploughman's spikenard, butterbur, yarrow, fleabane (*Peter Wakeley*)

20. Greater knapweed (*Peter Wakeley*)
    Carline thistle (*Michael Proctor*)
    Woolly thistle (*Peter Turner*)

21. Ramsons (*Peter Wakeley*)
    Cuckoo-pint (*Michael Proctor*)
    Flowering rush, arrowhead (*John Markham / J. Allan Cash*)

22. Birdsnest orchid (*Peter Wakeley*)
    White helleborine (*Michael Proctor*)
    Autumn lady's tresses, bee orchid (*Peter Wakeley*)

23. Crested hair grass, tor grass (*Michael Proctor*)
    Timothy, rough stalked meadow grass (*Mustograph*)

24. Marram grass, hare's tail (*John Markham / J. Allan Cash*)
    Greater pond sedge (*Michael Proctor*)

PREFACE

In this book I have tried to write an introduction to the vegetation of Great Britain with some emphasis on the lowland country, for after all, this is where the greater part of the people live. The plants described or mentioned are mainly common so that the reader can expect to find the majority quite readily.

I would like to thank my friend Mr R. Clarke for reading the original typescript and making very helpful suggestions. The Latin names have been quoted once in each chapter to encourage their use as they are more precise and scientific than the English names. This nomenclature nearly always follows that of the standard British Flora by Clapham Tutin and Warburg (2nd Edition 1962).

Farleigh,   C. T. PRIME
Warlingham,
Surrey

CHAPTER I

COLLECTION AND IDENTIFICATION

In order to gain some knowledge of plants, even if only to learn their names, it is necessary to collect a few specimens for examination. Everyone knows the names of some common plants that he or she has either seen or been told about so frequently that their names have become imprinted in the memory. Some may be common and conspicuous wayside weeds like buttercups or daisies while others like blackberry and strawberry may have attracted attention by their edible fruits or by their brilliantly coloured berries. With these plants there will be others, familiar perhaps, but not known so well that they can be named with certainty. These are the plants to start collecting for by this practice the collection of rare plants will be avoided. Take a country walk and gather only the plants which you can see are there in plenty. A few of these will keep you busy for quite a time and when you have mastered them you can go on to some of the remainder that are less common.

Make the collection with some care. If you just tear the flower away, you will almost certainly not come home with enough to make a proper identification and your efforts will be wasted. Try to decide if the plant is an annual, biennial or perennial, i.e. whether it lives for one, two or many years. Then collect a leaf or two, one from the base and one from the upper part and a single flower and a single fruit; in other words get representative parts of the plant so that from the pieces you can reconstruct a picture of the whole. Take written notes if this helps and if you do not feel able to take much of the specimen away. Sometimes it may be necessary to visit the plant more than once in order to get what you require. Willows for example, flower before the leaves emerge, and here you may want both leaf and flower to be certain of your identification. In others, quite a time may elapse between flowering and fruiting so a second visit may be necessary to get a fruit. When there are many specimens to choose from try to collect an average specimen, not the largest or the smallest. Put your specimens in a plastic bag (you can get really large ones) for this will keep them fresh until you have time to examine them.

By working in this way you will avoid the dangers of collecting too much and so reducing the frequency of our native plants. It should hardly ever be necessary to uproot a wild plant and the practice is to be strongly discouraged. Many common wild plants like primroses and sweet violets have almost completely disappeared from the neighbourhood of large towns and cities because people dig them up for their own gardens. Rare plants, also, have been greatly diminished in numbers by too frequent collection. No one has any right to endanger the existence of any plant and this rule of conduct is always to be remembered. Leave what you have seen for others to enjoy and learn to share your pleasure.

Different kinds of plant grow in different places. Keep your collections from different places separate from one another. In this way you will learn something about the habitats of plants. Woodland plants like the yellow deadnettle don't grow in the open and bog plants like the sundew are virtually never found anywhere else but in bogs. You will do well to seek out different places to visit in your neighbourhood. Try to find woods, heaths, rivers and marshes to visit in turn. Local guides, written accounts and maps of your district will often suggest places to visit, and if there is a local flora of your county this will always give

FIG. I. *Left,* snowdrop (*Galanthus nivalis*); *right,* aconite (*Eranthis hyemalis*).

you a lead as to places to visit and places to search for particular plants. But there is no substitute for sharp eyes and thorough searching. Visit your chosen places more than once a year, for some of the early spring plants have shrivelled away by early summer. Aconites and snowdrops flower in January or February, but they leave no traces above ground by June. Also visit your chosen places year by year for some plants do not flower every year. Bee orchids are notoriously variable in their appearance from year to year, and in some years there may be a blaze of colour from scarlet poppies by the wayside, yet in others, hardly any.

Only by practice and experience will you learn the sort of places to look for plants. Knowledge of the underlying geology and the type of soil often helps. Searching for soils derived from sandy deposits, soils derived from limestone and so forth will often lead you to good localities and in time you will get to expect to see particular plants from the general appearance of the ground. However field botany always has its surprises even for the most experienced, and therein lies some of its charm and continuing fascination.

Let us assume that you have your first collection at home and you wish to find out the names of your plants. First of all sort them out into their different kinds or species as they are called. For some plants this is easy, but not always for others. Kingcups and buttercups for example are easy to distinguish for they differ in size and leaf shape and in many other ways. The different common buttercups are a little harder but creeping buttercup has runners while the others have not, some have the sepals of the flowers turned downwards while others do not and there are other differences of a smaller order between them. However, some plants are more difficult; beginners often have difficulties with the yellow dandelion-like flowers of the Compositae and the white parsley-like flowers of the Umbelliferae. In sorting out the different kinds try to look for constant differences that are always there, rather than the sometimes more spectacular differences like variation in height and size. The constant differences may be small, for example, beech leaves have a wavy edge while hornbeam leaves have a toothed edge. Expect to find more than one of these constant differences, and look for them in the leaves, the stem, the flower and the fruit. Creeping buttercup (*Ranunculus repens*) has runners, it has a furrowed stalk, and its sepals are not reflexed, and its leaves are divided but triangular in outline. Meadow buttercup (*R. acris*) lacks runners, has stems that are not furrowed and leaves that are much divided but which are five-sided in outline, and its sepals are not reflexed. These are good examples of what are called specific differences and it is most important to learn to look for the right

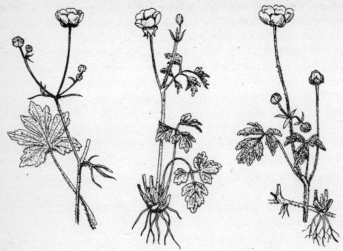

FIG. 2. Three species of buttercup: *left, Ranunculus acris; centre, R. bulbosus; right, R. repens.*

sort of thing. Sometimes you will find plants that differ in one character only; these are usually called varieties, for example a white-flowered bluebell or a cut-leaved beech.

Let us suppose that you have sorted out your plants into their different kinds and you wish to proceed to identify them with the aid of books. There are plenty available ranging from the simple and introductory to the complete scientific flora, and you must make your choice according to the state of your knowledge. However, beware of the limitations of any book that you choose, for nearly all of them describe only a selection of the flora, and in such a case you may come across something that is not in the book.

You may proceed to name the plant in one of three ways, by comparing your specimen with a drawing or photograph, by using a 'key' to 'run down' the species, or by comparing your specimen with a previously named and pressed specimen. The first method is a perfectly good way provided that the picture you use is a really good one and shows the botanical niceties by which the different species are distinguished. Be warned that some of the very small differences between say, the different species of speedwell may not show at all well in pictures and you can be led astray. Another reason for care with the use

of pictures is that there is no cheap book that gives you pictures of all
the wild plants you may meet in the British Isles.

You should therefore endeavour to use a flora for identification. In
these books identifications are carried out by means of artificial keys
by which species can be 'run down'. These keys ask you, in effect, to
put the plant you have in succession in one of two groups until you
finally come to the answer. Here is an example of part of a key taken
from a book in the New Naturalist series.

Key to the common hedgerow and roadside Umbellifers

1. Flowers yellow or pinkish   2
   Flowers white or pinkish   5
2. Leaves very much divided, segments hair-like, not all in
   one plane   *Foeniculum vulgare*
   Leaves 1- or more pinnate or ternote, leaflets flat, all in
   one plane   3
3. Leaves 1-pinnate   *Pastinaca sativa*
   Leaves 2–3-pinnate or ternate   4
4. Leaves 2–3-pinnate, leaves finely serrate   *Silaum silaus*
   Leaves 3-ternate, dark green and shiny, leaflets obtusely
   serrate or lobed   *Smyrnium olusatrum*
5. Stems spotted or blatched   6
   Stems unspotted   7
6. Stems glabrous   *Conium maculatum*
   Stems hairy   *Chaerophyllum temulentum*

You will notice that this and most keys use technical terms, and it is
necessary to learn exactly what they mean for it is these which give
precision to the key. Some you may guess; for example in separation 6,
glabrous may be guessed to mean without hairs since the other line of
the separation reads the opposite: stems hairy. But if there is any doubt
it is essential to consult the glossary. In separation 3, pinnate describes
a leaf consisting of more than three leaflets arranged in two rows along
a common stalk. Bipinnate means that the leaf is further divided and
that the stalks of the leaf are themselves pinnate. This latter example
shows how the use of technical terms saves both time and space, and
the effort required to learn and understand them is well worth making.

You can also identify your plant by reference to a named pressed or
dried specimen. Collection of these are known as herbaria, and many
museums, some schools and other institutions have such. Usually there
is more than one dried specimen of each so that the range of variation
of the species is shown, and so you can try to match your specimen
exactly.

Whichever method you use, start by checking the identifications of plants you know, and work the answer backwards through the keys of the flora. This will show how the keys work and it all helps to avoid mistakes. Then proceed to identify those plants about which you have some ideas but about which you are not certain, say for example a vetch or pea. You know it so far but are not sure whether it is a vetch or a pea, or what species of either it is. You will, however, be able to start with that key which separates vetches and peas, rather than starting right at the beginning of the book. Suppose you have a specimen about which you have no idea as to what it is. Then you have no option but to begin at the beginning to determine to which family the plant belongs. Species are grouped in genera and genera in families, the names of which usually end in -aceae and the first key in your flora should lead you to the right family. All this well may be difficult at first, but as you go, you will learn to recognise some of the families at sight. The appearance of most of the Cruciferae with their four separate petals in the form of a cross and very typical fruits makes this family very easy to know at sight, and so are the umbellate flowers of the Umbelliferae, and the pea-like flowers of the Leguminosae. By acquiring this perception which becomes almost an intuition, you will save yourself much time and as you go on you will get much quicker and more certain of your work.

Sometimes you will be quickly satisfied with your identification and feel quite confident that it is correct, but at other times you may have doubts. In such a case you will try all methods together. But it is as well to remember that there are some difficult groups of plants in the British Isles like the hawkweeds which can really only be satisfactorily named by experts, and the beginner is well advised to leave them alone.

Most of the books give the Latin names of plants and this practice has the great advantage that botanists the world over know what you mean by a name like *Ranunculus repens*. It has a precise meaning where other names can be vague. Although some people decline to use them, any person seriously interested should try to pick them up. Note that any plant usually has two names, first the generic and secondly the specific or trivial name. Thus all buttercups are put in the genus *Ranunculus* and each different kind is distinguished by an appropriate specific name. Thus the creeping buttercup is *R. repens* (repens=creeping) and the bulbous buttercup is *R. bulbosus*. Sometimes as in these two examples the names are clearly descriptive. *Ranunculus* itself is derived from rana (a frog) and is a reference to the fact that many buttercups grow in wet places. Often the meaning behind the name is more obscure. The

name of the pale hairy buttercup *R. sardous* means the Sardinian buttercup, and is an old historical reference to 'herba sardoa' a poisonous plant
thought to be a *Ranunculus*. The word *flammula* used for the lesser
spearwort, *Ranunculus flammula,* means a small flame and is perhaps a
reference to the burning taste of the leaves. Sometimes there is a third
name that describes the variety. A white flowered bluebell could
be described as var. *alba* and the weeping variety of the elm as var.
pendula.

The Latin names of plants are often followed by the name of the
person responsible for the plant name in the first case. Thus *Ranunculus
bulbosus L.* refers to Linnaeus, the authority for the name and who was
the man mainly responsible for the introduction of the present binomial
system of nomenclature. The naming of plants is subject to rules agreed
at International conferences and which are complicated to say the least,
and one result of them is that Latin names are subject to change. These
changes have been numerous in recent years and are a nuisance to
many. The best thing to do is to start with an up-to-date book and
learn the most recent names and hope they won't change, though at the
present time there does not seem much hope of this.

By contrast English names, if at times imprecise, have a constancy
that endures through the centuries. Daisies will always be daisies, and
names like heartsease, wallflower and milkwort are unlikely ever to be
forgotten. Like some of the Latin names they are often beautifully
descriptive; daisy is a corruption of day's eye, and dandelion is derived
from 'dents de lion' a reference to the toothed edges of the leaves.
Some names convey old medicinal uses like birthwort and herb bene't
(*Herba benedicta*='well spoken of'). Such names are part of our heritage
and always full of interest. Nor should it be forgotten that the gulf
between the Latin and the English names is not a wide one, for many
Latin names are in common use. People seem to prefer the Latin
antirrhinum to snapdragon and viola to pansy but perhaps it is only due
to the names on seedsmen's packets. In the West Indies, I am told that
Sempervivum is known as 'simple bible' and so in this case, as so often
in the remote past, the Latin is the source of the English name.

Records of the places you visit and the plants you find are always
worth making for their own sake and to aid your memory. The more
detailed they are the better; variations in the plants themselves are
worth noting and so are fluctuations in number from year to year.
Times of flowering particularly in the early part of the year always seem
to have a special appeal. In recording these, you will be following in the
steps of the greatest botanists for Theophrastus, often called the father

of botany writing about 350 B.C., records the following on February 1st at Athens:

> Violet, early bulbous (*Leucojum vernum*) in flower
> Wallflower (*Cheiranthus cheiri*) in flower
> Cornel (*Cornus mas*) in leaf
> Dogwood (*Thelycrania sanguinea*) in leaf

and later Linnaeus gives the following in the year 1755 for Sweden:

> Jan. 5 Rosemary (*Rosemarinus officinalis*) flowering
> Jan. 11 Honeysuckle (*Lonicera periclymenum*) in leaf
> Jan. 23 Red archangel (*Lamium purpureum*) in full bloom
> Hazel nut (*Corylus avellana*) in flower
> Laurustinus (*Viburnum tinus*) in full bloom

CHAPTER 2

THE PLANT COMMUNITY

ANYONE walking the countryside and looking for plants will be aware that some kinds of plants tend to occur together. Some species are found only in woodland company while others, like many sedges and rushes are only to be seen in wet places. One can go further than this and say that since the same groups of species always occur together in similar habitats, vegetation shows a degree of constancy of composition. Thus damp oak woodlands in the South of England all tend to contain very much the same species of plant and this can easily be tested by making lists for a number of woodlands and comparing them. Similarity in appearance and constancy of composition are the chief reasons that have led the ecologist to speak of vegetation as being made up of plant communities, and so we speak for example of the plant communities of heaths, of marshes and of bogs. But besides this 'constancy of composition' plant communities show other features. They show some degree of pattern or structure, for example, when they are organised in layers as in woodlands or heathlands or rivers. Thus there are free floating aquatics in rivers and ponds, submerged plants below the surface, and others that rise above the surface of the water. Even in vegetation of such a small height as grassland, the ground surface with its liverworts and mosses may constitute a distinct layer.

Furthermore another feature of a community is that there is often one species that dominates the remainder. A reed swamp around the edge of a lake often consists to such an extent of reedmace (*Typha latifolia*)

FIG. 3. A thallose liverwort, i.e. one lacking stems and leaves.

19

FIG. 4. A diagram showing the layers of vegetation in a mixed sedge community.

with smaller plants sometimes growing beneath it, that the reedmace is called the dominant; again, heathland almost completely covered with heather (*Calluna vulgaris*) which greatly affects other plants in its immediate neighbourhood may well be said to be dominated by it. Sometimes the dominance may be shared, as shown by the bent-fescue grassland of much of the British hillsides, while in many chalk downlands, the plant cover is so diverse that no one species can be said to be dominant.

This pattern or structure of a community is the result of the interdependence of the plants concerned. So often the dominant plant changes the climate near it so as to favour some species at the expense of others; it may cut down the light intensity to favour shade plants, or it may provide a sheltered spot for a plant that otherwise cannot withstand the full rigours of the climate. The tall upright shoots of reeds may slow down the flow of water in a stream so that the smaller free floating water weeds may flourish without being swept away, and the large roots may stabilise a muddy soil at their base that will allow small plants a foothold. The interdependence seen in plants growing above ground also occurs below the surface, where the root systems descend to different depths and exploit different layers of the soil. Bracken (*Pteridium aquilinum*) bluebell (*Endymion nonscriptus*) and wood sorrel (*Oxalis acetosella*) are plants that often occur together and have root systems that penetrate to different regions of the soil just as their shoots reach to different heights above the ground surface. Moreover, although green plants require the same mineral nutrients from the soil, some require more of one than another, and this is a further reason why some plants can flourish together without undue competition.

Unusually close associations of species with particular circumstances are found in parasites and saprophytes. Saprophytes like the birdsnest (*Monotropa* spp.) and the birdsnest orchid (*Neottia nidus-avis*) are dependent on a large supply of humus for their food materials, and so they only occur where there are large accumulations of dead leaves rotting to form leaf mould. An even closer association is found in parasites and semi-parasites for they are dependent on the presence of the right host plants. Some like the greater dodder (*Cuscuta europaea*) are virtually confined to one or two hosts, e.g. the stinging nettle (*Urtica dioica*) and the hop (*Humulus lupulus*), but others are not so particular and will grow on a greater variety of hosts.

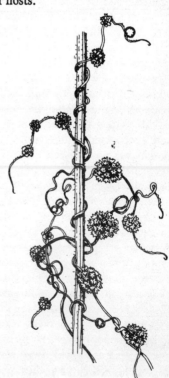

FIG. 5. Dodder (*Cuscuta europaea*).

Different kinds of plant communities may be found on different types of soil. The plants of chalk downland, are, as everybody knows, quite different to those of a sandy heath, and so are those of a salt marsh from those of sand dune. One element, calcium, makes a great difference to the plants a soil can support. Lime hating plants like the heathers and rhododendrons are described as calcifuge while the chalk-loving orchids are said to be calcicole. There are many other plants that are indifferent to the presence of calcium, while others are more particular in some areas than in others. Traveller's Joy (*Clematis vitalba*) is largely a calcicole in England, but it grows on a wider variety of soils on the Continent. As it reaches the northern limit of its range it can only survive on soils that suit it particularly well.

Climate also determines the type of plant community that any ground may carry; mountain tops do not grow trees, for they cannot withstand the force of the winds and there are many differences between the plant communities of the east and west sides of the British Isles that can be traced to the differences in climate between them. Blanket bogs develop only under conditions of high rainfall and high humidity, so there are none on the east side of the British Isles, but they occur more frequently as one moves towards the west of Ireland.

Climate is the sum of many factors including temperature, light intensity, light composition, light duration, rainfall, etc., and the elucidation of the separate and several effects of these is a very complex task. More than that, there are really many climates, for the climate inside a woodland is different to that outside, and the climate near the ground surface is different from that at a height of one or two metres. These smaller climates are called microclimates; they are obviously of great importance to really small plants like liverworts and mosses and to small animals like woodlice and others that are so liable to desiccation. Woodlice can flourish only on soil surfaces or wood crevices where the humidity is much greater than elsewhere.

Vegetation is profoundly influenced by the animals that live and feed upon it; in fact, many communities like downlands are maintained in their existing state by the animals that graze them. Man is the greatest influence of all, and as a result virtually all the vegetation of the British Isles is at best semi-natural, and most is completely unnatural owing to agriculture. The vegetation of mountain tops and perhaps of sand dunes and salt marshes comes nearest to the natural state and there are other small isolated areas which can reasonably be regarded as relics of the past vegetational cover of the country which for one reason or another have been left almost alone since earliest times. Wistman's wood on

Dartmoor is a few acres of oakwood growing on an area of huge boulders; when man cleared Dartmoor of its tree cover, this small area was left, no doubt, as being too difficult to clear.

Individual plants and animals vary greatly in their tolerance of all the variations that the factors of the habitat may provide. Some will only grow in a narrow range of conditions and are therefore limited in their overall distribution and in the communities among which they will grow while others will flourish over a very wide range. Plants like the common reed (*Phragmites communis*) are cosmopolitan occurring in almost every corner of the globe. Plant communities are also rather like this, some being more sensitive to environmental factors than others, and it is the least tolerant members that first begin to show signs of this; for example, the disappearance of less tolerant species in a pasture may be one of the first signs of overgrazing.

Plant communities are therefore not stable affairs; they are only kept in their present state by the action of one or many factors. Grasslands are stabilised by cutting or grazing and most of them would develop to some form of woodland if left to themselves. Ponds in the countryside tend to silt up and become dry land if not cleared out from time to time. Such changes are examples of plant succession, and the series of stages through which the vegetation passes during a succession is called a sere. Ultimately some form of climax vegetation is arrived at, and for most of the British Isles this would be deciduous woodland, a state to which it approximated in early historic time. Even the climax vegetation is not static, but dynamic, for within it plants are living, growing and dying all the time. The death of a large tree in a forest leaves a large gap that is filled in time by a process rather akin to the succession by which the forest was established in the first place. Such an event as the death of a large tree is very obvious, but careful observation shows that the same kind of thing occurs in all plant communities, so that the pattern, although apparently the same, is constantly changing.

There are a number of recognisable changes that occur during the course of a succession. In the early stages, the number of species increases greatly; this can be seen as a piece of waste land is colonised. These plants destroy their own habitat for they bring about changes in the environment so that it no longer suits them, and the original colonists are replaced by others; in the later stages the number of species may decline. A similar consideration holds good for the many associated animals for the number of species tends to increase as the succession proceeds and with this goes an increase in the complexity of their feeding relationships and of the interdependence of the animals.

So too, the production of total organic matter increases as the succession proceeds, but in the later stages it may well decline. These considerations have been shown to be true for forests, and they are among the most successful systems with a long history of survival, even over many geological periods.

The sun is the source of almost all the energy available to life on this planet, and the process by which it is made available to animals and man is photosynthesis. In this process, green plants take in carbon dioxide from the atmosphere and build it into sugars and other carbohydrates, storing light energy as chemical energy as they do it. Some of this energy is used by the plants for their own growth processes but most is stored and provides food for animals which set free the energy by their respiratory processes and use it for their own growth and development. Green plants are only able to absorb but a small fraction (1–5 per cent) of the light that falls upon them, and the animals that feed on the plants are only able again to use a small fraction (10 per cent) of the total energy they use as food. Thus there is a big energy loss at each stage in a food chain and this is the reason why there are usually not more than four to five stages in a chain. Simple food chains show three stages, such as plant →herbivore→carnivore, and an example might be:

Sheeps fescue grass→field vole→weasel

In nature the feeding relationships are never so simple for there are many organisms small and large that can feed on sheeps fescue and others besides weasels that can feed upon voles. Thus there is a very complex relationship between them, and the resulting situation is best described as a food web.

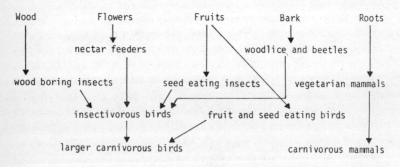

FIG. 6. A woodland food web.

Since food is used for energy as we have seen there will be a flow of energy through the whole system and a study of this flow is an important part of modern ecological study.

Man grows plants like wheat in serried ranks to obtain his food or perhaps more strictly, to obtain his supply of energy. Despite his care in the choice and selection of his crops the energy absorption of them is not usually equal to the best that nature can do, except where there is intensive all the year round agriculture, of which the sugar cane is perhaps the best example. Some regions of the earth's surface, like deserts, have a very low rate of primary production. Man's aim is to increase this and so to bring about increases in the food supplies of human beings the world over.

Since there is an energy loss at every step of a food chain most efficient use would be achieved by man if he were entirely vegetarian and fed directly on his plant crops. He does not do this but eats some animal protein and studies of how to produce this protein most efficiently are therefore important. These have led to the many present-day methods of animal husbandry such as the production of eggs by battery hens and the raising of calves and chickens in specially constructed houses. In most of these methods, advantage is taken of the fact that the young animal is more efficient in producing animal protein than the older. The particular kind of farming required to produce food in this way raises aesthetic and even moral questions, for some people regard such practices as quite wrong. However, as may be, enough has been written to show that the study of organisms and their energy relationships is relevant to what is perhaps man's greatest problem, namely, that of feeding the rapidly expanding population of the earth. So this diversion from the main theme of this chapter may be justified.

CHAPTER 3

TREES AND SHRUBS

TREES are the most conspicuous plants in the landscape; by their size and number they make up much of the countryside and they may determine and influence all the other plants that grow near them. They are objects of great beauty and much of the varied glory of the British scene is due to them. Moreover, trees have supplied many needs of Englishmen down the centuries from heavy timber for building ships and houses to wood for making furniture, as well as wood for fuel and many other uses. There is therefore every reason for the young student to start by learning to know trees. The most important species are common enough and easy of access, and collecting a small spray to name does no harm to the tree.

The two native species of oak must take pride of place since they are the most abundant trees in Britain. Almost everyone can recognise an oak by its characteristic leaves and acorns, by its rugged outline and kneed branches. The two native species are similar but the more important differences between them are given below and some are shown in Fig. 7.

The differences between the two species of oak

Character	Q. robur	Q. petraea
LEAVES		
Shape	Obovate	Ovate
Lobing	Deep, irregular 3–5(–6) pairs	Shallow, regular 5–6(–8) pairs
Leafstalk	Short ($\frac{1}{4}$")	Long ($\frac{1}{2}$–$\frac{3}{4}$")
Leafbase	Marked auricles	Tapering into stalk
Lower surface	Smooth (a few simple hairs)	Always some stellate hairs
ACORNS		
Colour (when ripe)	Pale fawn	Uniform dark brown
Longitudinal stripes (when ripe)	Olive-green on fresh mature acorns	Absent
ACORN STALK		
Length	1–3"	0–1"
TERMINAL BUDS	Small and rather blunt	Larger and more pointed

26

The common oak (*Quercus robur*) is the most frequent oak of the heavier and damper soils of the east of England while the durmast oak (*Quercus petraea*) is the most commonly occurring tree of the west side of Great Britain. This is only a rough approximation to the truth, for the durmast oak occurs on sandy soils in the east, and the common oak ·is found in the west. Where the two species meet hybrids are often found, and it is good practice to try to pick them out from their parents.

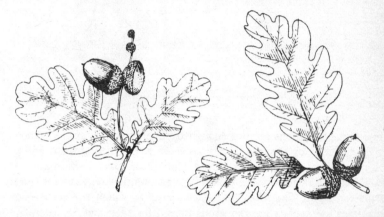

FIG. 7. The two common species of oak: *left, Quercus robur*; *right, Q. petraea.*

Oak trees form a substantial part of the British woodlands and are most common in the south.

Perhaps next in interest and importance is the beech (*Fagus sylvatica*) which is most frequently seen on the chalk lands and Chiltern hills of the south of England. It is a large tree with wide spreading boughs and it is readily known by its oval straight-veined leaves that have a wavy edge and which are fringed with very fine white hairs when young. In winter, it may be known by the long narrow pointed buds clothed with many overlapping scales. It bears crops of nuts which are really triangular fruits enclosed in bristly four-cleft cups. If you open the fruits many will be found to be empty and to be devoid of a seed. This is because the tree only produces good crops of seed every few years or so, in the so-called mast years. Actually the tree requires a high summer temperature to lay down an abundance of flower rudiments for the following season as in the hot summers of 1949 and 1959. Thus a fine flowering is the result of favourable conditions in the preceding year.

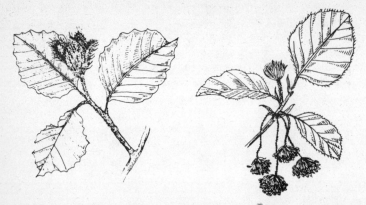

FIG. 8. Beech (*Fagus sylvatica*): *left*, female shoot; *right*, male catkins.

But these conditions are not enough by themselves to ensure a good yield of seed, for the tree is very sensitive to late spring frosts and a temperature of − 1°C during May will kill off the flowers. Even then, a moderately warm and moist period in late summer is required for good fruit formation, and after that the hazards to which the seeds and seedlings are subject are numerous indeed for they may be eaten by many small animals or be attacked by disease. The small rodents that eat the fruits are very much influenced by the winter temperature; a cold winter means few but a mild winter many. So although the fine summer of 1959 favoured flower formation, and the weather of 1960 allowed a mast year, the mild winter of 1960–1 meant a large rodent population and so the seeds and seedlings resulting from the heavy mast were almost all eaten. Thus do the factors affecting seed formation and development interact in quite a complicated manner.

A tree that resembles the beech in some respects is the hornbeam (*Carpinus betulus*). Its leaves are similar in shape but they have toothed margins and the buds though narrow and pointed are not nearly as long as those of the beech. Also they do not stand out from the twig as much as those of the beech. The trunk is a light gray in colour and often oval in section; in old age it becomes very fluted and develops characteristic buttresses. Its fruit is quite different in appearance from that of the beech for it is a small brown nut partly enclosed by unequal green wings by which it is readily dispersed. The tree occurs mainly in the south and east of England, often in the company of oak and beech. Its name is a reference to the hardness and toughness of the timber which was formerly used, amongst other things, for the cogs of mill machinery, for pulleys and for making wood screws. The tree is hardly ever so large as the beech (although Surrey has one over 100 ft in height) and it is slower growing. It is less frost tender than the beech, and it is able to grow well in heavy soils that the beech dislikes. In some natural forests on the Continent it forms a lower storey beneath the oak, rather as hazel does in Great Britain.

FIG. 9. Hornbeam (*Carpinus betulus*).

Ash (*Fraxinus excelsior*) as a tree is quite different from those that have been mentioned so far. For one thing the leaves are borne in pairs and divided into leaflets, and the winter buds are dome shaped and black in colour. Its leaf canopy is not dense and it casts little shade. It is one of the last trees to unfold its leaves in spring and as might be expected, it is susceptible to spring frosts. However, it usually sets seed freely enough which curiously does not germinate until more than a year after it has fallen. Though a common tree in the landscape and often a constituent of oak and beech woodlands it only becomes a dominant on some of the limestones where there is a good depth of soil and where it is not too dry. One of its frequent associates in these situations is the wych elm (*Ulmus glabra*) a quite different tree from the common elm. All elms have broad oval toothed leaves with asymmetrical bases that are borne singly on the stems. Wych elm has the roughest leaves of all, but the twigs are smooth (hence the name U.

glabra) and some of the leaves have a tendency towards becoming three-cusped at the apex. Grown by itself the tree develops a large fan-shaped outline and it rarely forms suckers like the common elm. The fruits are thin and oval with broad papery wings by which they are readily dispersed.

The common elms (*Ulmus* spp.), for more than one species is recognised and their distinction is a matter of some difficulty, rarely form woodlands by themselves. Nevertheless, they are frequent trees in the landscape and the elm is perhaps the commonest hedgerow tree in southern England. The trees usually sucker very freely indeed and the twigs are often corky and the bark very rough. The tree flowers very early in the year, producing thousands of small clusters of reddish-purple flowers that, to the observant eye, change the colour of the tree in the early spring sunshine long before the leaves begin to unfold. The winged fruits produced in vast numbers are only fertile in exceptionally mild springs and seedlings are rare. The origin of many of the English elms is uncertain but many think that some were introduced in Roman times as forage plants; they are still often very frequent near villages.

Elms have recently been severely attacked by a fungus disease known as Dutch elm disease. The first sign of the disease is a yellowing of some of the leaves of a part of the tree which starts about June. Later more of the tree may be attacked, and in a severe case the whole tree may be dead by the autumn. Often the infection does not spread beyond one bough and since the infection does not last beyond the season a tree may well recover. The disease is caused by fungus (*Ceratocystis ulmi*) and it is spread by elm bark beetles (*Scolytus* spp.) which make characteristic channels beneath the bark. The fungus blocks some of the conducting tissues of the tree so the parts above become short of water and nutrients and so die. The disease was first identified in Great Britain in 1927, and it has fluctuated in severity ever since tending on the whole to be unimportant. The reasons for the fluctuations are unknown, but since 1968 the disease has greatly increased in severity and this country may lose as many as a quarter of its elm trees. There is no cure but it is important to remove dead trees to reduce the number of bark beetles that carry the disease. Chemical control of the beetle is now being tried, but this treatment is expensive and it has not yet been found to be completely successful.

Other trees that can form woodlands of short duration, are the birches (*Betula* spp). There are two common species (*B. pendula* and *pubescens*) both thriving well on light sandy or heathy soils. They are beautifully slender trees with more or less triangular leaves and the flowers are

FIG. 10. The two common species of birch: *left, Betula pendula* and *right,*
B. pubescens.

borne in catkins that later give rise to millions of very light wind-
dispersed fruits. The leaves of the two species are different, *B. pendula*
having an acuminate tip and a double row of teeth, whereas *B. pubescens*
has an acute tip and a single row of teeth. The twigs also differ, those
of *pendula* being glandular while those of *pubescens* are slightly hairy.
The seeds germinate readily enough and spring up in vast numbers
wherever conditions are suitable. Since this tree is not valuable as
timber it is not surprising that the forester regards it as a weed. Birches
commonly occur in oak and some beech woodlands where the soils are
light, but they are light demanding and cannot succeed in the shade of
mature trees. As they are short lived they disappear as the woodland
approaches maturity. In parts of southern England, where the Scots pine
(*Pinus sylvestris*) is found, birch and pine often develop together, but
since the pine casts a deep shade, the latter may remain as dominant
for a long time as in such circumstances it is not easily replaced.

In the south of England most of the pine has spread from trees introduced in recent historical times, but in Scotland there are still remains
of the original natural pine forests. The Scots pine has the needles in
pairs and bears small woody cones first greenish but later becoming
light brown in colour. The bark of a pine tree becomes reddish at about
the height of ten feet as the tree matures and this is a very good character
by which to recognise it. There are several other species of pine that
have been introduced to Great Britain and they are not always easily
distinguished so it is as well to make a careful inspection of any tree
before attempting to give it a name.

There is another tree that can form woodland in Great Britain and
that is the alder (*Alnus glutinosa*). This, however, only grows on waterlogged soils like marshes or fens, by the banks of rivers or streams and

FIG. 11. Alder (*Alnus glutinosa*).

similar places. Owing to the drainage and lowering of the soil water table that has gone on for centuries the places where alder woods of any size can be seen are now comparatively few but there must have been more extensive forests in past times. The alder is a distinctive and easily recognised tree; it has dark green oval leaves that are broadest above the middle and which are slightly sticky especially when young. But its most conspicuous features are the catkins; the males hang downwards and are reddish in appearance, while the females become woody forming false cones about one half to one inch in length. These persist for some time so that the catkins of two separate years may be seen together on one tree.

There are other native trees that are less common and which play minor parts in British woodlands and amongst them are the poplars and willows. The aspen (*Populus tremula*) with almost round leaves that are constantly in motion is a small tree of open woodlands where it prefers the damper sites. Other native poplars include the grey poplar (*Populus canescens*) with coarsely toothed leaves that are white underneath; it is a large tree with a smooth greyish white bark to its upper boughs but with a fissured bark lower down. The true black poplar (*Populus nigra*) is a rare tree of wet soils in the south and east of England. Its leaves are roughly triangular and finely toothed but the branches of the tree arch downwards in a very characteristic manner and are frequently covered with large bosses. The most frequently seen poplars in this country are introduced trees that never occur in our native woodlands. They include the Lombardy poplar (*Populus nigra* var. *italica*), the white poplar (*Populus alba*) and a large poplar of hybrid origin (*Populus canadensis* var. *serotina*) often wrongly called the black poplar.

Most of the willows are shrubs rather than tall trees, but the crack and white willows are big trees of riversides. The white willow (*Salix alba*) forms a tall tree with a narrow outline and with white undersides to its narrow leaves whereas the crack willow (*Salix fragilis*) forms a tree with more widely spreading branches. There are several shrubby willows to be found in marshy and fenny ground and that play a part in the development of fen to carr and woodland. The commonest is undoubtedly the sallow (*Salix cinerea*) with blackish twigs and narrow leaves about three times as long as broad. Goat willow (*Salix caprea*) has brownish twigs, large winter buds and leaves almost as broad as long; it grows on drier soils, even on chalk downs. The catkins of these willows are gathered and used as 'palm' on Palm Sunday in churches although they could hardly be more unlike a true palm. There are several other species of willow in Great Britain; altogether they form an interesting and difficult group of shrubs not always easily delimited from one another and they require some study to know well. Willows are dioecious, that is the male and female catkins are borne on separate individuals. Their numerous bright yellow male catkins and silvery females are very striking in early spring when they colour the whole tree from a distance long before the leaves are fully unfolded. The flowers of willows are pollinated by insects, and on bright sunny days large numbers of early bees can be seen flying above and around these shrubs.

The shrubs of downland like those that occur on chalk and limestone are numerous and important. There are the hawthorns too well known to need any description. Notice the plural for one of the two species, the midland hawthorn (*Crataegus laevigata*) is more frequent in woods than outside. It has leaves with less shallow lobes than the common thorn, but the most important distinction lies in the fewer and slightly larger fruits which usually have two stigmas as opposed to one in *Crataegus monogyna*.

There are other common thorny shrubs. Purging buckthorn (*Rhamnus catharticus*) is common on calcareous soils; it has comparatively few thorns but the rounded leaves are borne almost but not quite opposite to each other. The flowers are in greenish yellow clusters and give rise to black berries. On a chalk slope growing with it there may well be dogwood (*Thelycrania sanguinea*) which has similar shaped opposite leaves, but decidedly reddish twigs. It bears white flowers in clusters, that give rise to groups of black fruits. Dogwood is one of the shrubs that rapidly colonises chalk slopes if grazing is removed. Wayfaring tree (*Viburnum lantana*) has broadly oval and relatively large leaves with twigs that are downy and very pliant when bent between the fingers. The flowers are once again white, in flat panicles three to four inches across and give rise to a group of berries that are red at first but which later turn black.

A very pleasing shrub of hedgerows and calcareous soils is the spindle tree (*Euonymus europaeus*) which has bright green twigs in contrast to those of dogwood. The leaves are narrow, opposite, while the flowers are greenish white and borne in small clusters. The glory of this particular shrub is the fruit which is a lovely four-sided pink capsule that splits open to expose the bright orange seeds within. When this shrub fruits freely it is brilliant in autumn sunshine. There are other shrubs of calcareous soils that will catch the eye; for example privet (*Ligustrum vulgare*) with narrowly oval leaves, white flowers and black berries and the elder (*Sambucus nigra*) with divided leaves, a straggling appearance

FIG. 12. The fruits of **a,** privet (*Ligustrum vulgare*); **b,** spindle tree (*Euonymus europaeus*).

and large flat dropping clusters of berries. The whitebeam (*Sorbus aria*) has irregularly toothed leaves with white undersides, and as these are caught by the wind, the whole tree looks white in contrast to the green of its surroundings. It is a member of the Rosaceae, bearing white flowers in flattish panicles and reddish fruits. Its relatives include the rowan or mountain ash (*Sorbus aucuparia*) and the hawthorns, a relationship that is evident from a first sight of the flowers and fruit.

Other shrubs are more frequently seen in woodlands. Hazel (*Corylus avellana*) coppice forms the understorey in many oak woodlands. The catkins borne on brown twigs as early as January are one of its noticeable features while later the leaves unfold; they are hairy, round and toothed on the margins.

Other associates of oak woodland are maple and cherry; these are essentially trees and the cherry or gean (*Prunus avium*) can easily become a very large tree. It is readily known by its flowers and fruit, but the smooth reddish bark of the younger branches should be observed as well as the rather long toothed leaves which have tiny glands at their base. Maple (*Acer campestre*) has leaves with rounded lobes and

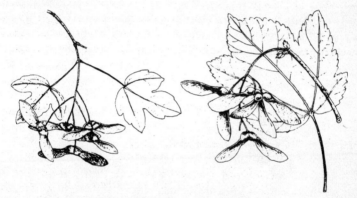

FIG. 13. *Left,* maple (*Acer campestre*); *right,* sycamore (*A. pseudoplatanus*).

its stems exude a milky sap; it flowers in April and May, producing a panicle of greenish yellow flowers. The fruit is well known with its two broad wings almost in a stright line. The sycamore (*Acer pseudoplatanus*) is a close relative of the maple and although not a native it is widespread and well established almost everywhere. It is hard to realise that it only arrived in the sixteenth century. This tree has leaves with sharp lobes and fruits like those of the maple but the angle between them is much less, about 60°–70°.

Occasionally in oak woodlands and hedgerows in the southern and midland parts of Great Britain, particularly in the weald of the south east one may come across the wild service tree (*Sorbus torminalis*). This tree is a relative of the whitebeam, but the leaves are deeply lobed with five to nine points. The flowers and fruit are again similar to those of the whitebeam but less conspicuous. The whole tree, however, turns a marvellous red in autumn, so much so, that despite its sometimes small size and rather infrequent occurrence, it can be picked out from amongst all the other autumn colours from quite a distance. Incidentally the presence of wild service tree is one of the signs of old relatively untouched woodland. Another very beautifully coloured shrub of autumn is the guelder rose (*Viburnum opulus*), which also has toothed and lobed leaves that turn scarlet in autumn. It bears a number of translucent red berries and it is a relation of the wayfaring tree.

Heaths and boggy places have their own complement of trees and shrubs. Birches are the most characteristic trees but there will be heathers and gorse as well as broom in the drier parts and also most likely bilberry. Common heather (*Calluna vulgaris*) is too well known to need description, but there are two other species one with purple flowers (*Erica cinerea*) and another with rose pink flowers (*Erica tetralix*). The Leguminosae is represented by three species of gorse, by broom and by three species of *Genista*. Petty whin (*Genista anglica*) is a spiny shrub that is thinly scattered on heaths and commons, while the second, dyer's greenweed (*Genista tinctoria*) lacks any spines and occurs in rough pastures. It formerly yielded a dye. The last, the hairy greenweed (*Genista pilosa*) is a rare plant distinguished from the others by its hairy twigs. Bilberry (*Vaccinium myrtillus*) is a very characteristic and easily recognised low shrub with green twigs, numerous small oval leaves and pink bell-shaped flowers that open very early in the year. It later produces the small succulent blue berries that make such excellent tarts. One other larger shrub worthy of mention is the alder-leaved buckthorn (*Frangula alnus*) a close relative of the buckthorn already mentioned earlier. This is a rather straggling shrub with flowers and fruit not unlike those of the purging blackthorn, but the berries first turn red and then later, a purple black. It occurs more frequently on peaty than on dry calcareous soils.

There are some evergreen trees and shrubs that play quite a part in the British woodland scene. The yew (*Taxus baccata*) is a frequent associate of the beech and it had many uses, such as for long bows, in earlier days. The red berries and narrow green leaves about an inch long as well as the reddish brown bark serve to identify it readily enough. More often than not, it has several trunks that tend to fuse as the tree grows older. A less common shrub is the juniper (*Juniperus communis*) with short needle-like leaves in whorls of three and with blue berries. Juniper is a declining species in the chalky south east of England, but it still holds its own in other parts of the country. The reason for its de-

FIG. 14. *Left,* yew (*Taxus baccata*); *right,* juniper (*Juniperus communis*).

cline is not known but it may be caused by disease or by pollution of the atmosphere or a subtle combination of these and other factors. Holly (*Ilex aquifolium*) is by far the commonest evergreen of woodlands but it is less common in the fens and on the clays than elsewhere. The flowers of any one tree are of only one sex, so only female trees bear berries.

There are a number of shrubby climbers like the honeysuckle (*Lonicera periclymenum*) that are found in woodlands, the latter especially on sandy soils. In woodlands honeysuckle does not flower freely and it apparently requires a certain intensity of light to do so. It forms part of the general undergrowth with such plants as the black berry (*Rubus fruticosus*). When the honeysuckle reaches the light, as at the tops of hedges it flowers very freely. The other woody climber is old man's beard (*Clematis vitalba*) with stems reaching twenty feet or more. It has leaves divided into leaflets and the plant attaches itself to supports by the leafstalks. The bark is very noticeable in that it is shed with age. The white flowers followed by the feathery fruits make this climber very conspicuous for it festoons quite tall trees and shrubs, particularly on calcareous soils.

There are about 150 shrubs and trees to be seen in Great Britain that are native or have some claim to have gained a permanent foothold here. Clearly, there are many others that have not been mentioned here and a keen observer will soon come across some of them. To identify them, reference must be made to larger books than this. Trees and shrubs make up the dominant part of the plant life of the country and while looking at the ground for smaller plants with bright and beautiful flowers, it is easy not to look up at the trees. What a lot is missed thereby!

CHAPTER 4

GRASSES, SEDGES AND RUSHES

BRIGHTLY coloured flowers attract the eye, and in consequence they
are often the first interest of the enthusiastic and botanically minded
observer. In searching for new plants he may seek out the brilliant and
elusive rarity and so pass by the plant that lacks a conspicuous flower.
Relatively large and brightly coloured flowers are easily recognised and
remembered, whereas smaller and inconspicuous flowers are harder to
learn and retain in the memory. Though nearly everyone starts by
learning the easy plants, it is better not to stop at that but to advance to
the more difficult, for the mastery of them gives much greater satis-
faction in the long run. Among the more difficult groups of plants many
people put the grasses, sedges and rushes and although this is true up
to a point it really is not difficult to learn many of the more common
members of these groups.

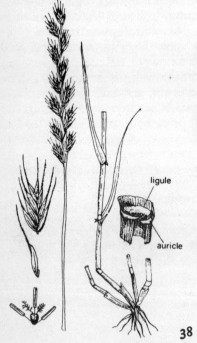

ligule

auricle

FIG. 15. Italian rye grass (*Lolium
italicum*). Notice the ligule and
the auricles.

First of all it is essential to distinguish between the three groups. Grasses have round or oval hollow stems that bear flat narrow leaves singly on opposite sides of the stem. These leaves have a sheath that surrounds the stem for a length before joining it at a swollen joint called the node. In grasses these nodes are more or less equally spaced. At the point where the leaf blade meets the stem there is often a small out-growth like a tiny collar of tissue paper. This is the ligule and it is often of importance in distinguishing species, for it can be long or short, pointed or blunt, ragged or smooth and so forth. Also at the point where the leaf blade meets the stem there may be a pair of auricles which take the form of small tongue-like growths going round towards the back of the stem.

The small flowers of grasses are arranged in the same way as the leaves, namely on opposite sides of an axis, as if in two vertical rows. The flowers are usually massed together in a spike or panicle of some kind or other, and normally though not always, each flower has both stamens and ovaries. The stamens are delicate and easily detached for they move freely in a light breeze and thus the pollen is carried by the wind to the stigmas which are branched and feathery. In some grasses where the flowers are truly small, these details are hard to see with the naked eye, but it a good pocket lens is used there is no difficulty if the flower is gathered at the right stage. The essential organs of the flower, the three stamens and the ovary and stigma are enclosed by two green scales; the lower is called the lemma and the upper the palea, and all these parts together constitute a single flower or floret. A number of florets are usually grouped together on a short axis with two further scales at the base called the glumes. Both the lemma palea and the glume may have fine drawn-out points known as awns. The florets and glumes together

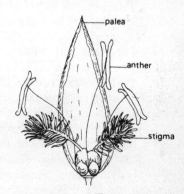

FIG. 16. The grass flower.

make up the spikelet, and the many spikelets are further arranged in spikes, panicles or clusters of one kind or another. There are many variants of this basic plan, and these variations go to differentiate the 150 or so grasses that have been recognised as native to Great Britain.

Secondly, the group known as the sedges are usually but not invariably perennials growing in damp or wet places. The stems are usually solid and often triangular in section. In some sedges, this triangular character is very obvious but in others the stem is only slightly three angled and then the character can easily be overlooked. A few have completely round stems. The leaves are long and narrow, indeed grass-like, but they can show a three-ranked arrangement related to the three-angled stem. They have a sheathing base like the leaves of grasses, but the sheath is never split. The leaves join the stem at a node, but in sedges there is always a long internode below the inflorescence; in other words the nodes are not so evenly spaced as in the grasses. In some members of the family the leaves can be very reduced, and then the green stem takes over the photosynthetic function of the leaves.

In sedges the flowers may have both stamens and ovaries or they may be unisexual. They are borne in the axil of a single scale (not between two as in the grasses) and arranged in one to many-flowered spikes or spikelets. Unlike the grasses the flowers are spirally arranged. Sometimes the male and female flowers are in separate spikes (again unlike the majority of grasses) and this makes it easy to tell the two families. However there is not a great difficult here even when the flowers are hermaphrodite and a little practice with the aid of a few pictures will soon enable anyone to separate these families. Usually at the base of each of the spikes there is a leaf-like bract which is often quite long and which does not occur in grasses.

Thirdly, there are the rushes which like many of the sedges are plants of damp places. The rushes are often tufted plants with a very erect habit of growth, and with leaves that are long and narrow and often round or nearly so in cross section. The woodrushes have flat grass-like leaves but these are readily known by the fringe of long white hairs along the leaf edges. The flowers of rushes are small, greenish or brownish in clusters or heads and each is readily seen with a lens to be star-like with six parts to the perianth and, usually, six stamens; in the centre there is a small ovary with three brush-like stigmas. It is not hard to imagine these flowers blown up to a greater size, when they would resemble a star of bethlehem or similar member of the Liliaceae a family to which the rushes are related, although this relationship is not perhaps apparent at first glance.

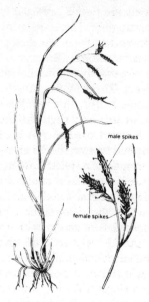

FIG. 17. Wood sedge (*Carex sylvatica*). Notice the terminal male and two or more female spikelets.

FIG. 18. A rush, *Juncus tenuis*.

There are a few other rush or reed-like plants that may cause some difficulty; for example, the bulrushes (*Typha* spp.) or more correctly the reedmaces which belong to a separate family of their own. Another small family is represented by the burr-reeds (*Sparganium* spp). which are very distinct by virtue of their spherical fruits and erect narrow leaves.

In order to know some grasses take a stroll round a garden and look for the weeds in a flower bed or on a path. Most likely the first grass you will see will be the annual meadow grass (*Poa annua*) which grows in open places almost everywhere in the British Isles. This grass has flattened non-hairy shoots with leaves that are often transversely wrinkled and which bear a ligule that is about 2–3 mm long. The flowers are in an open panicle and each spikelet is flattened and they may be found in flower at any time of the year. Another abundant grass is cocksfoot (*Dactylis glomerata*) a stout tufted perennial with very flattened shoots and tall stems bearing its crowded spikelets at the ends of the branches, the lower being the larger. It is frequent in grassland

and in pastures and on roadsides throughout the county. Rye grass (*Lolium perenne*) is a grass frequently planted where a hard-wearing grass is required. It has flattened shoots with sheaths that are reddish at the base. The ligule is blunt and there are tiny auricles where the blade meets the sheath that clasps the stem. The spikelets are flattened and arranged in an unbranched spike and placed edgeways on to the axis. In contrast to this is couch grass (*Agropyron repens*) with rather similar flattened spikelets placed edgeways on to the axis. It has abundant underground creeping stems which are such a nuisance to the gardener. Two relatives of couch grass (*Agropyron junceiforme* and *pungens*) and readily recognised as such, are to be found near the sea coast on dunes and in the drier parts of salt marshes.

Each habitat has its own characteristic grass species. Hedgerow and woodland, heath and moorland, chalk down, marsh and fen all have species peculiar to them. Thus on the chalk, a tall grass growing two to three feet high with broad leaves fringed with long hairs, and bright yellow stamens hanging from its open flowers will be the erect brome grass (*Bromus erectus*). Fine-leaved grasses from heaths and moors will include some bents and fescues. Flote grass (*Glyceria* spp.) is to be found lining the edge of ponds with its flaccid leaves actually lying on the surface of the water. Woodland grass include melic (*Melica uniflora*) the only British grass with a square stem as well as the wood poa (*Poa nemoralis*) with narrow leaves standing out all the way up the main stem.

Sedges show preferences like the grasses, though they are more frequent in damper habitats. Chalk downs will nearly always yield Carnation sedge (*Carex flacca*) which may be recognised by its leafy rosettes which suggest the colour of the foliage of the carnation. The inflorescence seen in early summer is quite different from that of any grass growing nearby.

Oak woodlands will usually show a green sedge. *Carex sylvatica* with quite long pendulous shoots 18 in. to 2 ft high. This is not to be confused with the pendulous sedge (*Carex pendula*) which is common in damp woods, but which is a much taller plant growing four feet or so high, with long hanging spikes looking like tassels. There are other members of the genus *Carex,* but they do not lend themselves to brief description since they are a very similar group and their separation depends on very many small points of detail.

There are other genera of the Sedge family Cyperaceae besides the major genus *Carex*; of these the cotton grasses are easiest to recognise by virtue of the long white hairs that lengthen so much after flowering.

The common rushes are again fairly selective in their choice of habitat. Heath rush (*Juncus squarrosus*) for example forms dense low tufts on moors, bogs and moist heaths. It has stiff channelled leaves with erect flowering stems with chestnut brown coloured flowers at their tips. It prefers acid soils where it is common, but it rarely occurs away from them.

The two commonest perennial rushes are the hard rush (*Juncus inflexus*) and the soft rush (*Juncus effusus*). Both form large tufts, two or three feet or more in height and are abundant in a wide range of habitats. In both the leaves are reduced to sheaths only visible at the base of the upright round stems. Hard rush has a slight bluish green appearance but soft rush has bright green stems; while the former is stiff and hard, the latter is soft and pliant. The flowers are in clusters towards the top of the stems, and are very similar, but careful study with a lens will show small differences in the shape of the capsules.

A rush of wet ground which does not show quite the erect habit of the two already mentioned is the jointed rush (*Juncus articulatus*) which grows a foot or so in height. The leaves of this rush have transverse partitions across that can be felt when the leaf is passed between the fingers. The little fruits are blackish and shining, darker than the rest of the flowers. It is not the only rush with partitions in the leaves though it is the commonest. There are also two distinct rushes that are almost confined to the seaside, the sea rush (*Juncus maritimus*) growing one to two feet high and usually to be found at the higher edges of salt marshes and a smaller species the mud rush (*Junus gerardi*) common in the salt marsh itself though rare elsewhere. More rare and only in the south is the sharp rush (*Juncus acutus*) known by its large size (1–6 ft in height) and its very prickly tussocks.

There is one common annual rush, the toad rush (*Juncus bufonius*) which is about 2 to 6 in. high and which usually grows on mud; it is slender and grass-like in its parts and it completely lacks the stiffness of

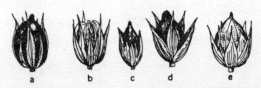

FIG. 19. The fruits of **a,** hard rush (*Juncus inflexus*); **b,** soft rush (*J. effusus*); **c,** jointed rush (*J. articulatus*); **d,** hairy woodrush (*Luzula pilosa*); **e,** Forster's woodrush (*L. forsteri*).

the rest of the genus. The flowers are small, pale green in colour and it is common in damp woodland rides, muddy places by ponds and in similar situations.

The woodrushes are, as mentioned before, distinguished by the grass-like leaves fringed with white hairs. The field woodrush (*Luzula campestris*) flowers from March to May and is therefore the earliest of these plants to be found in flower. It is common on grassy and heathy places throughout the country, and its flowers are just about the height of the turf in which it is growing. There are usually about three clusters each of about three to six flowers which show conspicuous yellow anthers at the time of flowering and which make a contrast to the chestnut brown flower segments. Many-headed woodrush (*Luzula multiflora*) is a rather taller plant that is to be found on acid and peaty soil, and it has about five clusters each with about eight to ten flowers arranged in a compact head.

The hairy woodrush (*Luzula pilosa*) is more frequently seen in woods and hedgebanks. It grows to about the same height as the many-headed woodrush, but the flowers are borne singly and arranged in spreading panicles. More abundant in the north and west of Great Britain is the greater woodrush (*Luzula sylvatica*) which is a far more robust plant forming bright green mats or tussocks. The leaves are about half an inch wide, which is more than in any of the other woodrushes and the flowering stem is also taller and larger in all its parts. It grows in woods on acid soils, sometimes in open moorland and often by rocky streams, but in the south it usually keeps to the shelter of woodlands.

FIG. 20. A single flower of the greater woodrush (*Luzula sylvatica*).

CHAPTER 5

THE PLANTS OF OAK WOODLANDS

OAKWOODS are the commonest woodlands in Great Britain, and the two kinds of oak are very similar trees and both provide a sheltered and comparatively well lit woodland in which numerous plants of various kinds can flourish. All woodlands show 'stratification' that is, the plants are arranged in tiers or layers, and there may be as many as four of these more or less distinct according to the type of woodland or the treatment it has received. They are the tree layer, the shrub layer, the field layer mostly of flowering plants, and the ground layer mainly of mosses that carpet the ground here and there.

Needless to say, oak woodlands vary greatly from place to place. The soil can vary from a heavy wet clay to a dry sand, the aspect can vary from a warm south-facing slope to a windswept slope facing north; one woodland may be managed by regular felling, replanting and coppicing while another may be left to itself. In any single wood there may be several distinct habitats. Woodland paths (rides) and clearings have a higher light intensity; stream sides and hollows provide damp habitats, and the presence of other trees besides the oak may alter the stratification and so affect the field layer. Therefore in the habitat itself there is plenty of variety, but, broadly, the woodland provides a sheltered habitat with a milder and less extreme climate than outside, and with a light intensity that varies with the degree of leaf canopy above.

Woodland plants vary in their tolerance of these conditions; there are some, but only a few, that you may find as frequently outside the wood as inside. Bracken fern (*Pteridium aquilinum*) is an example. Others may be so adapted that you never see them outside a woodland. Birdsnest orchid (*Neottia nidus-avis*) is such a plant being confined to woodlands where there is a sufficient depth of soil and leafy humus in which it can flourish. Between these two extremes there are all sorts, and in this chapter, there is space to mention some of the more frequent, and to say something of their features.

Woodland should be examined early in the year, for many of the plants are growing slowly all through the winter except when the temperature is very low, and their leaves are seen above ground in the first months of the year. The roughly circular leaves of the lesser celandine

45

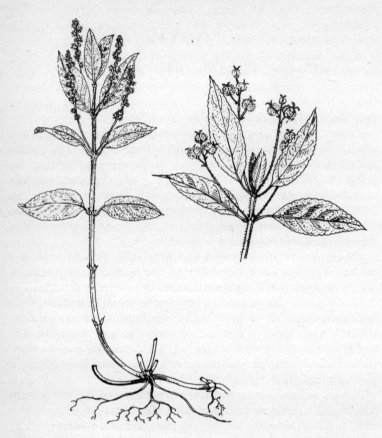

FIG. 21. Dog's mercury (*Mercurialis perennis*): *left,* male and *right,* female shoot.

(*Ranunculus ficaria*) appear in January, the arrow-shaped leaves of the common cuckoo-pint (*Arum maculatum*) in February and the shoots of dog's mercury (*Mercurialis perennis*) a little later in the same month. When there are so few other fresh shoots coming through the ground, the shape and appearance of these early plants identifies them at once, but if there is any doubt, they can be kept in view until they flower. It is wrong to think of winter as an unfavourable season for the growth of these plants; more likely unless it is very cold indeed, it is the dry summer period, for many woodland species spend this time in a completely dormant state. The wood anemone (*Anemone nemorosa*) is another early plant known by its single white or pink flower borne on a stem bearing three divided leaves or bracts. This is a character shared by all anemones. (You may have noticed the cluster of green leaves below the heads of the anemones you can buy in the shops for house decoration.) Wood sorrel (*Oxalis acetosella*) is another white-flowered woodland plant of early spring but this has smaller leaves each consisting of three undivided heart-shaped segments.

A smaller plant to look for is the moschatel (*Adoxa moschatellina*) which is usually found on the shady and damper soils. This has a rather pale green foliage and grows about 2 in. or more high; the leaves are divided into three further divided leaflets. The flower stems are un-branched and bear five small light green flowers, one on top and four below arranged in the manner of the faces of a town hall clock. In fact, 'town hall clock' is one of its common names. It is an interesting little plant; it has no close relations, it rarely sets seed, it dies away early in summer, and is easily overlooked. Oak woodlands of course, provide a number of very well known and familiar plants. The sheets of bluebell (*Endymion nonscriptus*), never seen to greater perfection anywhere else in the world cannot be excelled for beauty; primroses (*Primula vulgaris*), so abundant and full of flower in the woodlands of the west country are also a special pride of Britain. Less conspicuous are the violets, but they are worth looking for and studying. The first to flower is the sweet violet (*Viola odorata*); this has, as its name says, scented flowers that are normally deep violet or white; behind the petals there are green sepals with blunt points. This plant also has runners and is perhaps more common on hedgebanks, and plantations than in true woodlands. One violet like it, is the hairy violet (*Viola hirta*) but it is scentless, has no runners and is found on calcareous slopes and open woods. The leaves of this plant greatly enlarge after flowering, and are again, as the name says covered with long spreading hairs. The two most characteristic wood violets are somewhat later flowering and very similar to one an-other. Both have a central leafy shoot or rosette, but the flowers are borne on branches from this rosette; the commonest is *V. riviniana* with slaty blue flowers each having a pale spur at the back of them. The other wood violet is *V. reichenbachiana*; this has lilac-coloured flowers with a darker spur. The set and proportions of the petals in the two is different; in *riviniana* they form a rough square (hence the name 'old square face' while in *reichenbachiana* they are more oblong.

Buttercups are not very common in woodlands, but the creeping buttercup (*Ranunculus repens*) may be found in many wet places, particularly along the paths. Its long runners and divided leaves give it away at once. Lesser spearwort (*Ranunculus flammula*) may turn up in similar situations but it is more frequent on wet places on heaths. It has narrow spear-shaped leaves. In woodlands and roadsides and on heavy soils goldilocks (*Ranunculus auricomus*) is the buttercup to look for. This plant has basal roundish leaves, but the upper stem leaves are finely divided into long segments. The petals are golden yellow, but very often the flowers are incomplete and look as if the petals have been pecked or torn off. Actually research has shown that the species consists of many 'races' which set seed automatically without fertilisation occurring. Pollination is however only necessary as a stimulus to start the process off, and such being the case, hybridisation between the races is not possible.

As spring passes into summer many more species can be seen, but the great glory of the spring aspect has gone by. Among the labiates, which are an easily recognised group, the yellow deadnettle (*Galeobdolon luteum*) is conspicuous; it is the only common large-flowered yellow labiate, and it only flowers freely if the light intensity is relatively high. Other Labiatae of the woodland include the bugle (*Ajuga reptans*) with upright spikes of blue flowers and many creeping runners, and the common ground ivy (*Glechoma hederacea*), a plant that creeps and carries its flowers on one side of ascending stems; its flowers are blue-violet and show purple spots on the lower lip.

There are not many white umbellifers in the woodland, but two are important; one has many finely divided leaves and sends up single stems about nine to twelve inches high; this is the earth or pig nut (*Conopodium denudatum*) so called because of the small tuber it has underground. The basal leaves wither away at the time of flowering and the stem leaves soon after. It is not confined to woodlands for it occurs on heaths and in fields. The other white umbellifer, wood sanicle (*Sanicula europaea*) is known by its 3–5-lobed basal leaves which have toothed margins and its tiny fruits which are covered with hooked bristles and consequently can catch in your clothing.

Rides or woodland pathways have more light than the deeper woodland and some plants gain an advantage thereby. If it is really wide, then the side of the pathway approximates more to a hedgerow, and the plants are less typically those of woodland. The central path will have to withstand a certain amount of treading or cutting, so some smaller plants stand a better chance in consequence. On the floor may be seen creeping Jenny (*Lysimachia nummularia*) with stems bearing roundish leaves in pairs and yellow flowers. It roots very freely as it grows. By the sides of the path avens or herb bene't (*Geum urbanum*) may be seen; it is a yellow-flowered plant growing to twelve inches or more with a leaf made up of two to three pairs and with fruits that each have a single hook by which they may be dispersed. A related plant the wood aven (*Geum rivale*) has reddish purple flowers and is found occasionally in damp woodlands. Another plant with hooked fruits that occurs in woodlands is the enchanter's nightshade (*Circaea lutetiana*) which has small white flowers in a long spike and the fruits are covered with many tiny hooks rather than the single hook of the avens.

There is a group of yellow-flowered plants, the St John's worts that you find in or near woodlands. They all have opposite leaves, and if you hold them up to the light you can see clear spots (glands) in them, and in addition sometimes the sepals of the flowers are also edged with glands. Most characteristic of woodlands are the slender St John's wort (*Hypericum pulchrum*) which grows neatly about a foot or more high and is a very erect plant with roundish leaves and many pellucid glands, and by contrast, the trailing St John's wort (*Hypericum humifusum*) with pellucid glands only in the upper leaves.

In dry oakwoods, associated with *Quercus petraea*, besides many of the plants already mentioned, there are others which are very regular in their appearance. Woodsage (*Teucrium scorodonia*), a labiate with a spike of pale yellowish-green flowers is a common plant, and foxglove (*Digitalis purpurea*) too familiar to need description, is often found in the lighter parts of the woodland, where its spikes of purple pink or white flowers make a fine display of colour. Others, too, that occur in this type of woodland include the tormentil (*Potentilla erecta*) a yellow-flowered plant with four petals to each flower and with most of the leaves divided into three leaflets.

Woodrushes can be found in various kinds of woodlands and also on heaths and commons. They are grass-like plants with small star-shaped flowers in clusters. They all have grass-like leaves which if held up to the light and examined closely will show a few long silvery hairs along the edges. If the flowers are open, they will show quite clearly six star-shaped petals and they are really quite unlike the flowers of the true grasses. Of the species most likely to be met with in woods, the commonest is the hairy woodrush (*Luzula pilosa*) which grows to about nine to twelve inches in height and bears a panicle of chestnut brown flowers. The fruit is a small capsule which contracts above the middle to a little conical top. A very similar but less common southern woodrush is Forster's woodrush (*Luzula forsteri*) but in this plant the clusters of flowers are one-sided and the capsules taper more gradually to their tips. Where the two species grow together, a hybrid between them is often found. This may be recognised by the fact that it is infertile and also by its rather larger size.

The true grasses are represented by the millet grass (*Milium effusum*) which is about two to four feet high with flat wide leaves and a lax spreading panicle which has spreading or turned down branches. Another woodland grass is soft grass (*Holcus mollis*), a greyish white grass often spreading over quite large areas of ground and easily recognised by a ring of hairs at the nodes, and of course, there are some grasses like the meadow grasses which may turn up almost anywhere and will certainly be seen.

The once traditional system of forest management for oakwoods is known as coppice-with-standards. This means that the tree planting was so arranged that the oaks were equally spaced to about twelve per acre. Under the oak canopy a coppice of hazel and shrubs was allowed to develop and this was cut at intervals of ten to twenty years. This old system of management provided the oak trees and curved timber necessary for building ships and houses, while the coppice provided small wood for a large variety of uses from bean poles, hurdles and fencing, to charcoal for smelting iron ore.

There seems little doubt that the original oak forest of much of the country was something like coppice-with-standards, but the trees would have been closer together and shown a greater diversity of species, and there would also have been a similar greater diversity in the shrubs of the undergrowth. There still exist parts of forests like this in parts of the country that may never have been clear felled. The traditional management has left us with very beautiful woodlands for the spacing of the trees meant that they were open and well illuminated and thus a fine carpet of flowers could develop. This illumination falls gradually as the coppice becomes thicker and so the growth and vigour of the plants of the field layers diminishes. When, however, coppicing takes place and the shrubs are cut, there follows a large increase in the light intensity and a corresponding upsurge in the flora. This again diminishes as the coppice grows up and cuts down the light supply. Thus the full glories of a bluebell oakwood in the spring are really dependent on this old forestry practice which allows the necessary light for the development of the field and ground flora.

Unfortunately this traditional forestry practice is no longer com-mercially profitable, for the short trunks and crooked timbers it pro-duces are just what the timber trade of today does not require. There is no demand for coppice for charcoal and only a limited market for other uses. So for any landowner to continue the old forestry methods is to suffer financial loss and fewer landowners are able or willing to withstand this. There is a market for the best chestnut coppice in the south of England, where it is used for making hurdles and fencing, but to get the necessary straight and quick growth the chestnut must be grown as pure coppice and not under the shade of the oak. On many, if not nearly all, the soils of Great Britain the planting of oak woods is just not a financial proposition. Oakwoods can provide a wonderful amenity and most beautiful scenery, and anyone who does plant them deserves the thanks of all. The alternative is to plant softwoods like Corsican pine or Norway spruce which do pay to grow and a slow change to these kinds of woodland is inevitable. They can be beautiful and they can provide good amenity, but many people feel that many conifers are out of place in the English landscape. Wise planting often provides borders and avenues of hardwoods to break the monotony of the continuous conifers, but the effect is not the same as a large deciduous woodland.

Wide rides are also left in the extensive new forestry plantings not only for access but to act as fire breaks. These provide habitats for a variety of species like the heath milkwort (*Polygala serpyllifolia*), some ferns (*Dryopteris* spp.) and even occasionally a club moss (*Lycopodium clavatum*).

The majority of conifers grown are evergreen and therefore there can be no changes in the light intensity such as occur in deciduous woodland and which we have seen are so essential for the growth of field and ground layers that are so magnificent in our oak woodlands. Moreover,

FIG. 22. Stagshorn clubmoss (*Lycopodium clavatum*).

as well as being evergreen, the coniferous trees are planted so close together that the light below hardly allows anything to grow beneath them at all, so these woods will not become so interesting to the naturalist. It follows therefore that if we want to keep oak woodlands they will have to be conserved by deliberate effort. Areas must be set aside where the traditional practice of coppicing can be continued so that the rare plants of these woodlands such as the true oxlip (*Primula elatior*) of the boulder clay woods of East Anglia can be seen in all their beauty.

At the same time efforts must be directed to making woodlands that are managed for timber production as suitable as possible for the conservation of all forms of wild life. This means keeping as wide a variety of trees as possible by planting some hardwoods amongst the conifers and vice versa; it also means keeping existing small groups of trees when planting up whole areas. By so doing, it is possible also to maintain a diversity of structure and to provide some thick woodland, some thin woodland, some with a well developed understorey, and also plenty of paths and rides that suit some plants and animals particularly well. What has been said applies with greater force to the animals of woodland, many of which such as rare insects, require special habitats like rotting wood, old trees and particularly thick undergrowth which are usually not provided.

CHAPTER 6

BEECHWOODS

BEECHWOODS occur on chalk, limestone, sandy and even gravelly soils in the southern half of England. The southern slopes of the North and South Downs, the Chilterns and the limestones of the Cotswolds provide the warm dry and well aerated soils the tree prefers. The beech does not flourish in damp or waterlogged soils, and although it will grow well as far north as Aberdeen when planted, it appears to be native only in the south. John Ray did not include it in his list of Cambridge-shire plants published in 1660, and so in that county the few beech-woods must have been planted since his time. As mentioned in Chapter 3 the flowering and fruiting of the tree are much affected by late spring frosts and these may be an important factor in limiting its northerly distribution in the British Isles. Alternatively, it may still be slowly spreading and may not, as yet, have reached its climatic limit in this country.

On steep chalk slopes, the rain water washes the soil particles down the hillsides and the soils are very shallow. As the water drains away rapidly, the soils are dry and inhospitable. Beech grows in such places, and the soil beneath the trees becomes covered with a layer of leaves that decay very slowly. The slow decay is partly due to the dryness of the soil, and also to the absence of much ground vegetation, which exposes the leaves to the winds that sweep up the hillside. Where the slopes are gentle and where through drainage from above and around the soils remains damp, as on the plateau of the Chiltern hills, the soils are deeper and more like true loams. Here earthworms are more abundant, so the leaves are readily incorporated into the humus and a more fertile soil results. On sands and gravels as at Burnham beeches in Buckingham-shire, the nutriment is so readily washed out of the sand and gravel that a very infertile acid soil deficient in soluble mineral salts is formed.

The beech itself casts a dense shade; the main branches are flat, almost in one plane and the numerous smaller ones overlap and inter-twine in such a way that the maximum possible leaf surface is exposed to the light. This, of course, greatly reduces the light reaching the ground below, and this factor together with the poor soil, means that the plants of the field layer are restricted in number and kind. There is virtually no shrub layer, but yew (*Taxus baccata*) and holly (*Ilex aquifolium*) occur now and then and in the beechwoods on loams, and the blackberry (*Rubus fruticosus*) is a most important plant sometimes forming a continuous cover. It is, however, best considered as part of the field layer, since it grows and competes with the other plants of this layer.

A characteristic but not very common evergreen shrub is the spurge laurel (*Daphne laureola*). It grows about three to four feet high and bears clusters of small greenish yellow flowers in the axils of the leaves in February and March. The fruits are black and very poisonous. It is a close relation of the mezereon (*Daphne mezereum*) which is very frequently grown in gardens; the latter is a small deciduous shrub which is covered with pink purple flowers early in the year. It is a native of British ash and beechwoods on the limestone, though a very rare one. Another poisonous plant of scrub, perhaps rather than the pure beechwood, is the deadly nightshade (*Atropa belladonna*) which grows about three to four feet high, and bears dingy purple bell-shaped flowers followed by deceptive luscious looking black berries. It is famous for the drug atropine which dilates the pupils of the eyes. The stinking hellebore or setterwort (*Helleborus foetidus*) is a rarity that is to be found occasionally; its fresh light green spring foliage contrasts sharply with the much darker leaves of the previous season and its spray of greenish yellow flowers make it a most conspicuous and attractive plant in very early spring.

The thick mat of leaves on the floor of the wood is not very suitable for the germination and subsequent development of seed, and this is one of the reasons for the absence of a complete plant covering. Most of plants like dog's mercury (*Mercurialis perennis*) that do succeed in carpeting the floor of a beechwood have very good methods of spreading by creeping rootstocks or underground stems rather than by seed. The driest of beechwoods contain sanicle (*Sanicula europaea*) as one of the commonest herbs, and there may be ivy (*Hedera helix*) growing in the deepest shade. Other species include plants like the early wood and the hairy violets (*Viola reichenbachiana* and *hirta*), woodruff (*Galium odoratum*) and the wall lettuce (*Lactuca muralis*). Where the soil is deeper and conditions a little more favourable, sanicle gives way to dog's mercury and the number of different kinds of flowering plants increases. Woof (anemone (*Anemone nemorosa*), wild strawberry (*Fragaria vesca*), cuckoo-pint (*Arum maculatum*) and wood spurge (*Euphorbia amygdaloides*) are amongst the commoner species that may be found.

The thick mat of leaves and humus are specially favourable for the growth of two very interesting plants, the birdsnest orchid (*Neottia nidus-avis*) and the birdsnest (*Monotropa hypopitys*). These plants absorb the foodstuffs they require directly from the humus and are to be compared in this respect with the moulds and other fungi which live on the dead leaves in the woodland. Some of the most characteristic species of the beechwood are some members of the orchid family known as the helleborines. The broad-leaved (*Epipactis helleborine*), the violet (*Epipactis purpurata*) and the white (*Cephalanthera damasonium*) all occur more frequently in beechwoods than elsewhere.

On the relatively deep soils of the plateaux of the Chilterns blackberry becomes important together with many of the plants already mentioned. Other species like wood sorrel (*Oxalis acetosella*), yellow deadnettle (*Galeobdolon luteum*), the male fern (*Dryopteris filix–mas*) will also certainly be found, and woodland grasses like millet grass (*Milium effusum*) tall fescue (*Festuca arundinacea*) and occasionally wood barley (*Hordelymus europaeus*) become more frequent.

Beechwoods are less frequent on sandy and gravelly soils, and gravel seems to be the limit as far as soil is concerned. The trunks of the trees in such situations are often crooked and do not grow to the normal height and the ground flora consists of a very different group of plants to those seen in the chalk beechwood. Bilberry (*Vaccinium myrtillus*) and bracken (*Pteridium aquilinum*) are frequent, together with soft grass (*Holcus mollis*) and wavy hair grass (*Deschampsia flexuosa*) and there is no sign of sanicle or dog's mercury. Plants like the violets, the hairy woodrush (*Luzula pilosa*) and wood sage (*Teucrium scorodonia*) are not infrequent.

From time to time trees of the beechwood fall and die leaving a large gap and the resulting increase in the light intensity leads to the growth first of annuals and perennials, then shrubs, and later saplings of ash (*Fraxinus excelsior*), oak (*Quercus robur*) and others. Great competition follows between the saplings as the gap is closed and over the years regeneration of the canopy becomes complete with the growth of the young beech trees which overtop their competitors. The details of the development differ according to the size of the gap, the age of the woodland as a whole, and the time at which the gap occurs in relation to the occurrence of a mast year. Such a development is called a reproduction circle, and it is interesting to look for and to study some examples in woodland.

Beech trees, have comparatively few associated trees. One previously mentioned is the yew (*Taxus baccata*) and occasionally this forms areas of woodland by itself particularly on the sides and round the heads of chalk valleys or coombes as they are called in the south of England. One very famous example is at Kingley Vale near Chichester and has been described as 'the finest yew wood in Europe'. Yew invades scrub vegetation but does not colonise grassland direct, the scrub being necessary to give the developing yews some protection. Ash too, invades the scrub on the South Downs, but although it checks the yew development for a time it outlasts the ash and ultimately becomes the dominant. I do not think anybody is quite sure why in these particular places one ends up with yew and not beechwood. Rarely, as at Staffhurst wood in the weald of Surrey, a portion of old woodland has oak and beech thriving together; they are then said to be 'co-dominant'. The natural establishment of calcareous beechwoods has been studied extensively and in all its different habitats there are well-known stages through which the vegetation passes before the beech arrives to become the final dominant. On chalk grassland the sequence of the succession is:

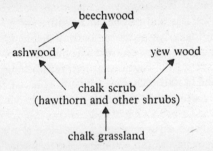

Thus the beech does not directly invade chalk grassland and therefore it is not surprising that the beech is not usually planted directly on such soils. The forester now forms plantations with conifers and young beech and the purpose of the conifers is to nurse and protect the beech trees in their early stages from the extreme effects of the climate. Young beech plants also suffer greatly from damage by rabbits and grey squirrels so their establishment is not always successful without care and vigilance. As the trees get larger the conifers can be removed and sold, leaving the beech trees themselves that can be sold when mature for profit. Many fine beech woodlands, like those at Goodwood in Sussex were established in this way with the aid of conifers as nurses. So the outlook for future beechwoods looks a little better than for oakwoods, but again, those with plants and animals of special interest will need to be 'conserved' if we are to keep them.

CHAPTER 7

CHALK DOWNLANDS

CHALK downland is widespread in the south and east of England; the North and South Downs run through Kent, Surrey, Hants and Sussex. There are further large areas of chalk downland in Wilts and Dorset while what used to be called the East Anglian heights run through Huntingdonshire, Bedfordshire and Cambridgeshire to reach the sea at Hunstanton. Chalk downland is one of the most enjoyable parts of the countryside to explore for it is pleasant to walk upon and much of it is high enough to give superb views of the surrounding countryside. Moreover, some of it is readily accessible as public open space and there are many tracks and pathways across it; for example, the South Downs Way in Sussex provides the basis of a splendid tour.

Most of the downs have been much as they are today for a very long time, almost certainly since the ways of Neolithic man who felled the trees and opened up the landscape. The downs have been maintained to comparatively recent times by sheep grazing aided by the ubiquitous rabbit which was introduced by the Normans. Now that sheep grazing does not play the part in agriculture that it used to and rabbits fluctuate in numbers according to the incidence of myxomatosis changes in the composition of chalk downland are to be seen. These changes are not always for the best for areas are liable to rapid invasion by shrubs like dogwood, hawthorn and others to the great detriment of the herbaceous plants the naturalist treasures so much.

The characteristic turf contains many grasses and a very large number of other herbaceous plants, all of which must be able to withstand grazing to a greater or lesser extent. Grasses have stems and leaves that are literally green right down to ground level, so that if the top part is bitten off there still remains some green part that can photosynthesise and grow. Moreover these grasses produce so many lateral shoots from the lowest parts of the plant that they spread over a wide area forming a turf. The grasses that form large clumps and stools are not found on downland for the green shoots would most likely be the first to be bitten away or trodden down.

Among the commonest grasses are the fescues (*Festuca* spp.) that are low tufted plants with very fine bristle-like leaves with brownish sheaths. Sheeps fescue (*Festuca ovina*) is the commonest with leaves less than 1/16 of an inch wide and a flower spike of about eight to twelve inches. Red fescue (*Festuca rubra*) is very like it but it is rather larger and has a more spreading habit with sheaths to the basal leaves that are reddish in character. These are common grasses of dry and poor soil and they are not in any way specially characteristic of chalk grassland. There are other grasses that only infrequently are seen away from chalk and limestone, like the two oat grasses; these may be recognised by the long awns that project from the spikelets. One, the meadow oat (*Helicto-trichon pratense*) is a rather stiff bluish green non-hairy plant, while the other (*H. pubescens*) is hairy, particularly so in the lower sheaths. Totter or quaking grass (*Briza media*) is another very easily recognised grass for its oval spikelets which shake so readily in the light wind make it unmistakable. It is especially characteristic of poor soils and it is often an indicator of old and unploughed grassland.

One very characteristic grass of the chalk downland is erect brome grass (*Bromus erectus*) which does best where there is relatively little grazing. Another grass the crested hair (*Koeleria cristata*) is a hairy perennial with a greyish green narrowly oblong panicle. Sedges are comparatively few on downland but carnation sedge (*Carex flacca*) is nearly always in evidence; it spreads by long underground stems and the typical rather purple-brown scales enveloping the yellow-green fruits give it away on examination.

Many other plants make up the short turf of downland but like the grasses they must be able to withstand grazing by some means or other. Some of them form leaf rosettes that lie very close to the surface of the ground. Good examples are the daisies and the plantains. The latter includes the ribwort (*Plantago lanceolata*) with narrow leaves and well-marked veins which is one of the very commonest of British plants. Chalk downs show another plantain, the hoary (*Plantago media*) with broader and downier leaves than the ribwort. It has a brighter and more attractive spike of pink flowers which are scented; this is because this plaintain is pollinated by insects in contrast to the other species which are wind pollinated. Other rosette-forming plants are some members of the Compositae like the rough hawkbit (*Leontodon hispidus*) with deeply toothed hairy leaves and yellow flower heads borne singly on the stems. The outer florets of the flower heads are orange or reddish on the back. The flower stem is also hairy and the plumed fruits that are left when the petals fall are a rather dirty brown, and each of the hairs of the fruit can be seen to be branched or feathery with the aid of a lens. Another rosette forming composite is the stemless thistle (*Cirsium acaulon*) in which not only are the leaves closely appressed to the ground but they are prickly into the bargain so the plant is doubly safe from the grazing animal.

Other plants grow close to the soil without forming a rosette as such but forming a mat of interwoven branches. Good examples of mat-forming plants are the common medick (*Medicago lupulina*), the birds-foot trefoil (*Lotus corniculatus*) and the horseshoe vetch (*Hippocrepis comosa*). Sometimes the branches of the mat root very freely indeed so

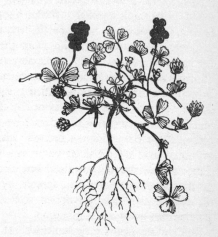

FIG. 23. Black medick (*Medicago lupulina*), a mat-forming plant.

that the whole plant is anchored down at many points, whereas in others there is only one central root, so on detaching this, the whole comes away. Mats and rosettes have other advantages besides avoiding grazing; they are less liable to damage from treading than if the individual leaves were erect and they are also placed at a very suitable angle to the incident light for photosynthesis.

Most of the downland plants are perennials and in such densely covered ground there are very few open spaces to be occupied by annuals. When a plant dies and a small area is left, it may well be occupied by a rapidly growing annual species like fairy flax (*Linum catharticum*). That this is so can be seen by carefully mapping a small area like a square metre or less over a period of years when it will be seen that although the overall pattern remains the same, different areas are occupied by different plants in successive years. One can envisage a situation where the percentage composition of any area might be the same in any two successive years. But if so, it would not be due to the same plants in the same places. It is this constant change that allows of bare ground from time to time and this is invaded by annual species.

There are, too, occasional small areas of downland on steep slopes where the soil is so shallow that plants grow poorly and bare areas occur. These sites are often sites for some of the rare chalk annuals like the ground pine (*Ajuga chamaepitys*), the cut-leaved germander (*Teucrium botrys*), the candytuft (*Iberis amara*) and, in Kent only, the rare milkwort (*Polygala amara*).

Downlands in summer are frequently hot and dry places and the water supply is far from the surface. Many of the downland plants meet this difficulty by rooting at depth, some like rock rose (*Helianthemum vulgare*), the dropwort (*Filipendula vulgaris*) and the salad burnet (*Poterium sanguisorba*) possessing root systems that penetrate 3 to 4 feet into the ground. Moreover the extent of the root system seems to be immense when it is compared with the small part of the plant that actually is to be seen above ground, for like an iceberg, nine-tenths or more is concealed below the surface. In addition, plants in dry places restrict their water loss in some ways or other. Here again, the rosette habit is valuable for the leaves pressed close to the ground diminish evaporation from the lower surfaces and the rolled leaves of grasses show another way by which the area exposed to the sun's evaporating power is diminished. The tuberous roots of plants may act as water storage organs and tide plants over periods of drought. Yet despite these adaptations, for some plants like the surface rooting grasses drying up is a danger, and in dry seasons, brown patches of grass may be seen, often in marked contrast to the green shoots of some of the deeper-rooting perennials.

Many of the plants of downland are common plants that will grow in any dry well-aerated soil and are really not very particular where they grow. This is true of the ribwort already mentioned, of the fescues, the birdsfoot trefoil, the meadow clover (*Trifolium pratense*) the yellow bedstraw (*Galium verum*) the knapweeds (*Centaurea* spp.) and a host of others. There are other plants that are almost confined to chalk down-land and very rarely found elsewhere. Such plants are said to be exclusive, and they mostly like the special chemical characters of the chalk soil that distinguish it from other soils. Chief among these are some of the chalk orchids, like the bee, the musk and the man. These are decreasing plants, partly owing to the scarcity of suitable places for them to grow, and largely owing to indiscriminate picking. They are short-lived perennials, often flowering but once in their lives and so if only one is picked, that plant has flowered to no purpose so far as the continuance of the species is concerned. If several people take but one plant each, the whole may well add up to a total devastation of that particular colony. It is therefore all important to avoid taking any specimens of our rare British plants.

The monkey orchid (*Orchis simia*) is now restricted to two localities in the country and the military orchid similarly (*Orchis militaris*) and both species are heavily protected. In case anyone is tempted to doubt the need for this, you have only to visit a protected nature reserve like one of those run by the County Naturalists Trusts, to see the beneficial effects of protection from over gathering, too much treading, and interference.

There is a host of other plants that can claim attention as being almost special to chalk and limestone downland. There is the pasque flower (*Anemone pulsatilla*) of the Hertfordshire and Bedfordshire downs, the round-headed rampion (*Phyteuma tenerum*) of the downs of Surrey and Sussex, the chalk scabious (*Scabiosa columbaria*), the field fleabane (*Erigeron acris*) and the felworts (*Gentianella amarella* and its relatives). The list is a long one. Some parasites and semi-parasites are also to be found in the chalk flora. Tall broomrape (*Orobanche elatior*) which parasitises the greater knapweed (*Centaurea scabiosa*) is confined to the chalk and one or two of the eyebrights which are semi-parasites of grasses are annuals that are also very much chalk species. The eyebrights are a difficult group and many introductory books are content to list only one inclusive species. However, *Euphrasia nemorosa* is a comparatively small flowered species of early summer which is widespread in many habitats including chalk downs, while *Euphrasia pseudokerneri* is a large-flowered species, flowering much later, that is more restricted to chalk downs. The latter plant shares with one of the felworts (*Gentianella anglica*) the distinction of being endemic to the British Isles, that is, they are not known elsewhere in the world. One other semi-parasite is worthy of mention, namely the bastard toadflax (*Thesium humifusum*) which is uncommon and confined to the chalk. This plant has a pale yellow-green foliage with narrow leaves and this somewhat distinctive colour helps to pick it out from among the grasses on which it grows. The flowers are small, pale greenish white and easily overlooked. It is the only representative of the sandalwood family to be found in England; most of this family, the Santalaceae are tropical.

This wonderful heritage of downland flora is very vulnerable, particularly so when more and more people seek out open spots of down for recreation. Take a careful look at a piece of chalk down that is regularly visited by many people. Quite apart from the losses caused by direct picking of some of the rarer species, there are other subtle changes. Where people tread most, the vegetation may disappear altogether but even where there is moderate treading the flora will be seen to have changed, for some of the chalk-loving species have disappeared and other plants have taken their place. This is due not only to the intensive treading but also to the fact that the soil has become enriched from the remains of food like fruit peelings and from the excreta of animals like dogs that the visitors have brought with them. In these circumstances grasses like the perennial rye grass and cocksfoot replace the chalk sward and the characteristic species disappear.

Cessation of grazing leads to the growth of bushes and if allowed to continue unchecked, it leads to the complete disappearance of the grassland. Also there may be more subtle changes such as the spread of tor grass (*Brachypodium pinnatum*) which, if it gets the chance, almost completely drives out the chalk plants. The young shoots of this grass are eaten immediately they come through the ground but if not eaten then, they become unpalatable and are left alone to achieve an almost complete dominance of the down.

It is clear then that only positive action will prevent undesirable changes taking place. The details of the necessary management will differ from place to place according to the species it is desired to conserve. On very thin soils of steep slopes no great treatment may be necessary beyond removing any bushes that start to grow. On more gradual slopes and deeper soils, the ideal practice would probably be to revert to a grazing regime with sheep at a number of approximately three per acre. This is not often practicable, and then the area must be mown to keep the turf down to the normal height and to prevent the growth of shrubs. The programme has, however, to be carefully planned. It is, for example, helpful to control tor grass by burning in winter, but excessive growth of erect brome grass is best checked by cutting it in April. The effects of undue treading can be avoided if paths are altered at intervals, and everybody should be encouraged to take home his own litter, not only for aesthetic reasons but for reasons of practical conservation.

CHAPTER 8

GRASSLANDS

CLOSE turf grazed by sheep and rabbits such as that of a down contains some grasses that tend to make much growth from the base of the plant; they are said to make much bottom growth. Meadows on the other hand that are cut for hay to provide winter feed for animals contain grasses that are taller and make much less growth at the base. Such grasses, too, make rapid growth in spring and early summer which is mown round about mid-June and again later if the season is favourable. Other grass on neutral soils is known as permanent pasture and really consist of grasses that need a fairly rich or well-treated soil to grow well. There is no hard and fast line between meadow and permanent pasture for the treatment to which they are subjected is the major cause of the differences between them.

Among the grasses of the meadow perennial rye grass (*Lolium perenne*) is one of the most important to the farmer for it is a highly nutritious food for cattle. With it may be associated cocksfoot (*Dactylis glomerata*) also highly nutritious and even more valuable because of its high yield. Others are timothy (*Phleum pratense*) and meadow foxtail (*Alopecurus pratensis*) and these both have cylindrical flower spikes. Timothy has the neater and more parallel-sided spikes while those of meadow foxtail taper more at the top and bottom. These grasses, together with cocksfoot are amongst the tallest and can grow to a height of four feet or so in a good hayfield. The meadow grasses, particularly rough stalked meadow grass (*Poa trivialis*), are also bound to be present; these have flattened shoots and the typical spreading panicles of the genus *Poa*. *Poa trivialis* has a long pointed ligule, but *Poa pratensis* has a short ligule, and more important, the former creeps over the surface of the ground by creeping stems, while the latter spreads by underground creeping stems.

Meadows always contain a number of other plants, some regarded as weeds, but others, like the clovers, are valuable for the nitrogen they accumulate in their root nodules. These are all members of the Leguminosae, and, agriculturally, the clovers are the most important. In meadows the common red clover is as important as any but there may be alsike clover (*Trifolium hybridum*) and white clover (*Trifolium repens*),

though the last named does not grow as tall as the others. Vetches like
the common tare (*Vicia sativa*) may be present and another frequently
to be seen is the meadow vetchling (*Lathyrus pratensis*). Common, too,
are the two species of birdsfoot trefoil, one rather low growing (*Lotus
corniculatus*) and the other the marsh birdsfoot trefoil (*Lotus uliginosus*)
which is hairier, and capable of growing to four feet or so amid supporting vegetation. It is helpful to remember when learning these plants
that clovers and medicks have trifoliate leaves, while the birdsfoot trefoils
have what look like five leaflets (really three leaflets and two large
stipules), that vetches usually have a large number of leaflets with
a tendril or tendril point, while the peas (*Lathyrus*) have a smaller number (two to four pairs) of leaflets with or without a tendril or tendril
point.

Numerous other flowering plants may be found in meadows; buttercups are amongst the first that come to mind and there are at least three
kinds: the meadow (*Ranunculus acris*), the creeping (*R. repens*) and the
bulbous (*R. bulbosus*) that one may expect to see. Buttercups are bitter-
tasting herbs avoided by animals, but when the plants are dried for hay
the bitter principle is destroyed. Not surprisingly they are absent from
the modern re-seeded and weed treated grasslands. A meadow plant
harmful to livestock is ragwort (*Senecio jacobaea*) but it is only really
common on poor soils. It acts as a cumulative poison. Other plants,
if not exactly poisonous, can cause trouble to the dairy farmer. Crow
garlic (*Allium vineale*) when eaten by cattle can give its flavour to the
milk and even yarrow (*Achillea millefolia*) can impart a bitter and aro-
matic flavour to milk and butter.

When the meadow is on damper soil, a number of interesting species
may well occur; for example the ragged robin (*Lychnis flos-cuculi*)
looking like a tall slender pink with ragged petals and the sneezewort
(*Achillea ptarmica*) a white-flowered member of the Compositae, and if
you are lucky some of the marsh orchids (*Dactylorchis incarnata*).
Such damper meadows are less likely to be ploughed and re-seeded than
meadows on drier soils and so their vegetation can be varied and
interesting. Of special interest are the water meadows lying in the valleys
of rivers and streams. These may be flooded in winter by heavy rains,
or deliberately submerged by means of holding sluices in the main
rivers. This, as done in the fen country, had the further advantage of
providing wonderful skating, should the winter be severe enough to
allow thick ice to form! Flooding brings a layer of alluvium to the
meadows that contain valuable minerals and so the resulting crop
benefits. Such a water meadow might contain some of the rushes and

PLATE I. Ranunculaceae: *above left*, pasque flower, a species confined to south-facing slopes of chalk downland; the leafy divided bracts are characteristic of the genus *Anemone*. *Right*, green hellebore, an uncommon plant of calcareous woodland which does not set seed very regularly. *Below left*, traveller's joy. The climbing habit, the opposite leaves and the hairy fruits contrast with the rest of the family. *Right*, wood anemone has very leafy bracts.

PLATE 2. Cruciferae: *above left*, garlic mustard, a common hedgerow and woodland species which favours basic soils. *Right*, scurvy grass, a plant of muddy shores and estuaries with fleshy leaves and globular fruits. *Below left*, candytuft is a rare crucifer of stony calcareous slopes, very bitter to the taste and thus avoided by animals. *Right*, cuckoo flower, common in damp meadows and pastures; the petals are in the form of a cross and the pods are long and narrow.

PLATE 3. Droseraceae: sundew. Confined to *Sphagnum* bogs, the plant catches insects by the glands on the leaves.

PLATE 4. *Above*, Hypericaceae: *left*, hairy St John's wort showing the hairs and the glands on the sepals which are a feature of this plant; grows mainly on basic soils, in woods and copses. *Right*, common St John's wort. The stem is round with two raised lines upon it; the leaves show many translucent dots when held up to the light. *Below*, Violaceae: *left*, wood violet. Note the relatively square shape of the flower and the spur which is notched at the apex. *Right*, yellow mountain violet. Though often yellow, the flowers vary in colour to blue or violet; it is found mainly in hilly districts.

PLATE 5. Caryophyllaceae: *above*, moss campion, a typical cushion plant of mountains, has a growth form well adapted to survive exposure. *Below*, stitchwort. The opposite, entire leaves and the floral parts in fives are very characteristic of the family.

PLATE 6. Leguminosae: *above left*, broom. The flowers are explosive and only visited once by the pollinating insect; note the unvisited flowers to the right and the others in which the stigma and stamens can be seen. *Right*, gorse, found usually on the lighter and less calcareous soils, especially if the ground is disturbed; the floral mechanism is similar to broom. *Below*, black medick with the small curled pods of the genus *Medicago* and the point to each of the leaflets.

PLATE 7. Rosaceae: *above*, barren strawberry, often confused with the common strawberry; note the spreading hairs, distant petals and small terminal tooth to the leaflet. *Below*, mountain avens has eight petals and a very characteristic leaf shape. The leaves of mountain ladies mantle can be seen in the foreground.

PLATE 8. Crassulaceae: navelwort is typical of crevices and walls; the petals are fused into a whitish-green tube.

PLATE 9. *Above left*, Saxifragaceae: purple saxifrage is a mat-forming plant of mountains, found both in the arctic and the alps. Crassulaceae: *right*, English stonecrop showing the fleshy leaves and characteristic inflorescence. *Below*, roseroot, a plant of mountain crevices, but found also on sea cliffs in the north and west of Britain.

PLATE 10. *Above left*, Cornaceae: dogwood, one of the commonest of chalk shrubs, rapidly colonising downland when grazing is removed, and very difficult to control. Caprifoliaceae: *right*, wayfaring tree, common in southern England, has rather thick opposite leaves and paniculate heads of flowers. *Below*, guelder rose showing the outer ring of sterile flowers, and the inner mass of fertile flowers in bud. The red berries and the autumn colouring of the leaves make it one of the most beautiful of our native shrubs.

PLATE 11. Umbelliferae: *above left*, hemlock water dropwort grows mainly by streams; not calcareous. A typical umbellifer with finely divided leaves and bracts and bracteoles beneath the branches of the umbels. *Right*, wood sanicle, a very important species of beechwoods and of oakwoods on loamy soils; the fruit is covered with bristles. *Below left*, parsnip, a roadside and waste-ground plant. The flowers are yellow and the leaves only once divided; the whole has a strong smell. *Right*, cow parsley, one of the commonest, and the first to flower, of the white umbellifers; to be found in hedgerows, the edges of woods and waste places.

PLATE 12. Ericaceae: *above left*, bell heather favours the drier heaths and moors. Note the clusters of small leaves and the bell-shaped flowers. *Above right*, heather is found only on acid soils where it can become dominant over great areas; also abundant in open woods with such soils. *Right*, cross-leaved heath is common on the wetter heaths and moors. It has waxy bell-shaped flowers and the leaves are in groups of four.

PLATE 13. *Above left*, Labiatae: yellow deadnettle, a woodland species, rarely flowering in very low light intensity; it prefers heavier rather than light soils. Scrophulariaceae: *right*, black mullein occurs on open banks and waysides, more frequently on calcareous soils. The flowers are in a long spike and are almost regular. *Below*, eyebright is a large-flowered eyebright, endemic to the chalk of southern England.

PLATE 14. Scrophulariaceae: foxglove occurs on acid soils throughout the British Isles. This picture shows the tubular flowers and characteristic capsules of the genus.

PLATE 15. Labiatae: *above left*, red deadnettle, a very common weed of waste ground. Note the grouping of the flowers, the opposite leaves and the shape of the separate flowers. *Right*, wild basil has a circle of finely divided bracts beneath the flowers. *Below left*, bugle, a woodland species with many creeping stems, a very short upper lip and a larger, lower, three-lobed lip. *Right*, cut-leaved germander, a rare labiate of south-east England; its distribution is southern continental, just reaching into England.

PLATE 16. Campanulaceae: *above left*, clustered bellflower, a plant of chalk grassland and similar calcareous places. Note the cluster of flowers, all without stalks. *Right*, nettle-leaved bellflower grows in woods and hedgebanks with perhaps a preference for chalky clay. Note the angled stem and large coarse teeth to the heartshaped leaves. *Below*, harebell. This photograph shows the fine teeth of the calyx and the delicate flower stem very clearly.

PLATE 17. *Above left*, white bryony is the only British member of the Cucurbitaceae; it has a massive tuber, and exceedingly quickly grows very long shoots which rapidly cover a hedge. *Right*, Rubiaceae: yellow bedstraw is common on all types of grassland. It has leaves in whorls and small flowers with four joined petals. *Below*, woodruff, a woodland plant with leaves in whorls, angled stems and four joined petals.

PLATE 18. *Above*, Dipsaceae: *left*, teasel is common in waste places, especially on clay soils. Like the Compositae, the flowers are massed in a head but the stamens are free. *Right*, scabious, a plant of waysides and pastures, differs from most Compositae in having opposite leaves. *Below*, Gentianaceae: *left*, perfoliate yellow-wort, a chalk-loving species of duneland. *Right*, felwort. Note the fringe at the throat of each flower, the number of petals and the form of the sepals.

PLATE 19. Compositae: *above left*, ploughman's spikenard. Note the shape of the flower heads, the narrow green bracts surrounding each head and the softly hairy leaves. *Right*, butterbur has large leaves and flower heads. This species is dioecious, and the photograph shows the male flowers; the female plant is much less common than the male. *Below left*, yarrow is common in grassland everywhere. It is readily known by its very finely divided leaves. *Right*, fleabane is common in damp places; it has both yellow disc and ray florets, heart-shaped stem leaves and a generally woolly appearance.

PLATE 20. Compositae: *above left*, greater knapweed, a plant of dry places. The outer florets are sterile, only the inner are functional. Note the shape of the bracts beneath the florets. *Right*, carline thistle. The persistent straw-tipped inner bracts spread outwards and look like the petals of a flower; this also occurs in the everlastings of our gardens. *Below*, woolly thistle. The woolly heads are most characteristic of this thistle; our British plant is an endemic subspecies.

PLATE 21. *Above left*, Liliaceae: ramsons has broad leaves and umbellate flowers; common in old damp woodlands. *Right*, Araceae: cuckoo-pint, a common hedgerow and woodland plant of basic soils, variable in most of its characters. *Below left*, Butomaceae: flowering rush, a rather local plant of rivers, ditches, ponds and canals. *Right*, Alismataceae: arrowhead grows in similar places to the flowering rush. The male flowers are at the top of the spike, and the female flowers below.

PLATE 22. Orchidaceae: *above left*, birdsnest orchid, a saprophyte, is dependent on the soil humus for all its nutrients. *Right*, white helleborine, typical of beechwood orchids. *Below left*, autumn lady's tresses, a late-flowering orchid of calcareous soils. *Right*, the well-known bee orchid is found on a variety of soils from chalk or limestone pastures down to chalky clays and calcareous dunes.

PLATE 23. Gramineae: *above left*, crested hair grass and *right*, tor grass are typical of chalk downland. *Below left*, timothy and *right*, rough stalked meadow grass are both meadow grasses.

PLATE 24. *Above*, Gramineae: marram grass holds sand dunes together with its numerous underground stems. *Below*, Cyperaceae: *left*, hare's tail is a cotton sedge of moorland, while *right*, the greater pond sedge may be found in swamp conditions.

FIG. 24. Fritillary (*Fritillaria meleagris*).

sedges, possibly some marsh orchids, the water speedwells, brooklime, water dropworts and others making altogether a most interesting habitat. One of the most fascinating plants of the British flora is the fritillary or snakeshead lily (*Fritillaria meleagris*) which grows in some of the alluvial meadows fringing the Thames in the Reading district. This plant is well worth making a special expedition to see.

Pasture is regularly grazed grassland and most of it that can be seen in southern England has been ploughed and re-seeded with grasses of high nutrient value to the animals that feed on them. The seed mixtures used contain high percentages of some of the grasses already mentioned like perennial rye grass, cocksfoot, timothy, together with white, red or alsike clover. It must be confessed that such pastures can be very dull for the naturalist and yield for him only the occasional weed or the unusual invading grass. The weeds will consist of plants that cattle leave alone, like beds of nettle (*Urtica dioica*), dock (*Rumex* spp.) and thistle. Chief among the latter will probably be the spear thistle

(*Cirsium vulgare*) with erect purple heads an inch or more across and large very spiny leaves. The two commonest docks are curled dock (*Rumex crispus*) with narrow leaves and wavy edges and broad-leaved dock (*R. obtusifolius*) with broader heart-shaped leaves at the base. Both can grow to a height of three to four feet and are avoided by cattle on account of their acid taste. None the less land is very variable and the depression here or the hillock there may provide some plant a little out of the ordinary even in a comparatively fresh and recently sown pasture.

Many of the hillsides in north west Britain, particularly where the soils are acid and generally poor, form rough hill grazing and the main grasses of which it is made are the bents (*Agrostis* spp.) and the sheeps fescue (*Festuca ovina*). These hillsides, like the rest of Great Britain were once covered with tree and since their clearance, the present state has been maintained by grazing. The area of grazing may reach as high as the arctic alpine zone of the higher mountains and extend down to the valley regions that it is possible to cultivate more intensively. Also *Agrostis*-fescue grassland is not by any means confined to the mountainous districts for it may occur in many other parts of the country like the heaths of East Anglia and the other parts of southern England.

The species of *Agrostis* of hill pastures are thin, fine-leaved grasses with soft foliage and they are not always easily distinguished. Brown top or common bent (*Agrostis tenuis*) is the most abundant; it has a rather tufted habit, grows to about a foot high and has brown spreading panicles which do not close up after flowering. The little ligule on the non-flowering shoots is blunt and shorter than broad and the plant spreads by underground shoots. The other abundant *Agrostis* is *A. stolonifera* (creeping bent) which, as its name suggests spreads by numerous creeping stems that go over the surface of the ground rooting freely at the nodes and thus producing new plants at intervals. In this grass the panicle closes after flowering and it is whitish rather than brown in appearance. *Agrostis canina* variously called velvet or brown bent has a panicle that is spreading in fruit and the flowers have tiny awns, which are readily visible with a lens. Sheeps fescue has already been described (p. 58) and it is a very common grass.

This grassland often has a number of associated species of which the commoner are the tormentil (*Potentilla erecta*) with four yellow petals and toothed divided leaves, the heath bedstraw (*Galium hercynicum*) a milkwort (*Polygala serpyllifolia*) and the field woodrush (*Luzula campestris*). Where the ground is damper rushes will be found and where springs come out of the hillside to form streams bright green patches of vegetation will be seen which are known as flushes. Among the plants growing there may be water blinks (*Montia fontana*), various mosses, one or two sedges and in the water itself a small water crowfoot like *Ranunculus omiophyllus* and one of the forget-me-nots (*Myosotis* spp.).

Agrostis-fescue grassland is liable to invasion by other species. Bracken (*Pteridium aquilinum*) is one of the most frequent invaders and an unwelcome one because it is left alone by grazing animals. The fronds may kill off the plants beneath them by cutting off their light supply and also by falling on them when they die back in autumn. Bracken spreads largely by its tough long underground rhizomes but at the same time it produces many much shorter laterals that bear most of the fronds. Not all the available buds come up in any given year so that it has many in reserve, so if the first to come up are damaged in any way, there may be a second emergence. The spread of bracken is limited by frost for although the fronds do not come through the ground till late May, they can still be damaged. The fronds too, are intolerant of exposure to strong winds, so that in mountain areas, bracken tends to be restricted to valley slopes. It will not grow in soils that are waterlogged in winter so it is often missing from the bottoms of many valleys in the hills.

Where the soil is more peaty and acid a grass known as mat grass (*Nardus stricta*) often covers large areas. It is also known as white bent because of the whitish hue of the older leaves as they become bleached by the light. It possesses creeping stems from which it sends up shoots and so it spreads. The leaves and sheaths contain much fibre and the grass is therefore useless for grazing. The flowering stems bear the flowers all to one side giving a toothbrush-like appearance to the flowering spike. Dog's tail (*Cynosurus cristatus*), a meadow pasture grass similarly has its flowers to one side, but the brush is much thicker; in the mat grass the brush is reduced to a few hairs, looking more like a comb with a few fine teeth. Mat grass is not confined to the northern parts of the country for it does occur in the lowlands though it is not nearly so common. It is favoured by over grazing, particularly in the wetter districts, because the rain washes the soluble minerals away impoverishing the soil. It has few associates but the heath rush (*Juncus squarrosus*) with its green rosettes and upright stems bearing clusters of flowers at the tips is characteristic.

Where the soil is wetter, purple heath grass (*Molinia coerulea*) is common. This is a large tufted grass with flat leaves and spreading panicles of flowers. The roots are unusual for a grass, being cord-like and penetrating deeply into the peat and soil on which the plant grows. It is found in almost all parts of the country, in the south as well as the north, but the reasons for its detailed distribution are not always clear. On Exmoor for example where it is extensive, it seems that *Molinia* has replaced areas once covered with heather that has disappeared, largely as a result of grazing. *Molinia* seems also able to tolerate a high iron supply and at the same time to have only a small lime requirement.

CHAPTER 9

HEATHS AND COMMONS

HEATHS are widespread in the British Isles, particularly in the south. We think of them as wide open areas largely covered with heather (*Calluna vulgaris*); almost certainly there will be parts covered with bracken (*Pteridium aquilinum*), some grassland and here and there, gorse (*Ulex europaeus*) and young trees of birch and others establishing themselves. Characteristically heaths occur on acid sandy soils like the greensands but they are to be found on other geological formations. The soil developed under these plants usually shows a thin black peaty layer about one inch thick, below which there is a lighter zone of almost whitish sand, to be followed by a blackish layer at a depth of nine to twelve inches. The white colour is due to the rain washing out soluble

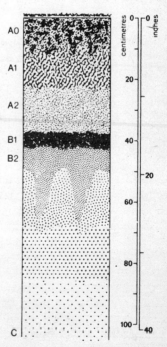

FIG. 25. A podsol profile. Ao = humus layer, A1 and A2 = bleached layers, B1 and B2 = layers of deposition and C = parent rock.

material from the upper layers and depositing it again at a greater depth. Such a soil profile, as it is called, can easily be seen if a fresh exposure of the surface is made by deliberate digging, or by breaking away the vertical surface at the edge of a sand pit.

Heather is the dominant plant of heaths and it is too well known to need any description, but in passing, one may note its very numerous small leaves closely packed together. This feature is to be interpreted as a means of cutting down the water loss during periods of exposure to desiccating winds. Clearly the water loss would be greater if the same total area of leaf surface was constructed as a few large flat leaves rather than as a series of imbricating shoots. Actually, heather cannot withstand long continued drought; it becomes much less frequent on the Continent as one goes eastwards and southwards, not liking either the extremes of a continental climate or the heat of the Mediterranean. Like a number of other plants, heather forms a mycorrhizal association with a particular fungus. The fungus lives in a zone of tissue in the roots and without it the plant cannot grow well; very likely it assists in the absorption of nutrients from the soil particularly where the soils are poor, and after all, it has to be remembered that heather is predominantly a plant of poor soils.

Heather, when growing well, seems able to resist invasion by other plants like bracken or tree seedlings. But the balance is a delicate one and comparatively slight changes will tip the balance against it. Heather does not stand up well to much treading; this may cause some branches to die and make spaces where the seedlings of other plants can grow. Fires too are a very important factor that can cause much damage; they may so destroy the heather that tree seedlings develop, but if fires are recurrent then the young trees are destroyed and a continuous cycle may be set up.

Associated with the heather there may be other species like the bell heather (*Erica cinerea*) which has dark green leaves, three in a whorl, and usually grows in drier areas than ordinary heather. In wetter areas will be found the cross-leaved heath (*Erica tetralix*) so called because the leaves are in groups of four, and they also bear glandular hairs. Common gorse (*Ulex europaeus*) is often found on heathland and its presence is greatly aided by any disturbance of the ground. It is a tall very prickly shrub, dark green in colour, which flowers in the early part of the year reaching a peak in April but which will often start as early as November and flower in mild spells through the winter. Though seeming hardy enough, it is killed back in really severe winters and this allows the heather to regain areas it may have lost to the gorse. There are two other species of gorse to be found in the south of England. One is the dwarf gorse (*Ulex minor*) that does not grow nearly as tall as the common gorse and which is a lighter green in colour and which also has paler yellow flowers. It is often to be found growing between and among heather and it flowers during the late summer, the yellow making a lovely contrast to the purples of the heaths and heathers. Rather more difficult to distinguish is the western gorse (*Ulex gallii*); it is a stiff spined plant, in many ways intermediate between the two, and it is predominantly to be seen in south west England and Wales and much less frequently elsewhere.

Gorse should not be confused with broom (*Sarothamnus scoparius*) which is not spiny and which has masses of light yellow flowers and which prefers the drier and the more sandy soils. Another yellow-flowered leguminous plant is the petty whin (*Genista anglica*), a small spiny shrub with oval leaves growing to a height of one to two feet. It bears pointed pods that disperse their seeds explosively. Though found here and there it never becomes a really important member of the heath community. Where the heath becomes damper or where there are the beginnings of a tiny valley, purple heath grass (*Molinia coerulea*) may well be found and such a valley may well mark the beginnings of a boggy patch. In drier places wavy hair grass (*Deschampsia flexuosa*) and the heath grass (*Sieglingia decumbens*) will be found together with the tormentil (*Potentilla erecta*) and the heath bedstraw (*Galium hercynicum*). In the west country there is a species of *Agrostis* (*A. setacea*) the bristle bent, that is abundant on many heaths. However, there are other special treasures of the west country. In the neighbourhood of Poole, the heaths blossom with the beautiful pink flowers of the Dorset heath (*Erica ciliaris*) which is known by its deep pink flowers and one sided spikes. Further west, the heaths of the Lizard also have a plant of their own, namely the Cornish heath (*Erica vagans*) which bears pale lilac flowers in quite dense heads. The Lizard is an area of the south west favoured with a mild climate and a special geology, so it is not surprising that its vegetation is unusual and that it includes a number of species not found elsewhere in the country. Elsewhere in the west country, the upland heaths grade into moors and the type of vegetation seen on Dartmoor, Exmoor and Bodmin moor.

On the east side of England, in East Anglia, a vegetation of great interest can be seen on the Breckland heaths. Though now largely afforested and including one of the largest planted forests in the country, there are still parts covered with heather and also parts covered with grass heath of one kind or another. The soil is varied and mostly

glacial in origin but it is nearly always sandy in character. In the nineteenth century much of the area was covered with heather which was
largely maintained by rabbit grazing. It was shown by Pickworth
Farrow, one of the early ecologists, that intensive grazing causes the
replacement of heather by grassland. On the sandy soils of the Breckland an important constituent of the grassland is the sand sedge (*Carex
arenaria*) and this is eaten by rabbits but only where other more
palatable species are not available. The final result of intensive grazing
leads to the development of a zonation round a rabbit warren, firstly,

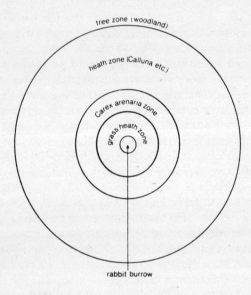

FIG. 26. Plant zonation surrounding a rabbit warren on a heath in
Breckland. After Farrow.

in the immediate neighbourhood of the burrows there is a zone of bare
soil in which would be found only plants like the nettle (*Urtica dioica*),
thistles and elder (*Sambucus nigra*) which rabbits will not eat; round this
there is a zone of very densely grazed turf, and it is interesting to note
that the turf owes its existence to the very grazing that is destroying
it all the time. Further away there is a zone of sand sedge (*Carex
arenaria* that the rabbits will, but prefer not, to eat and finally the
original heather which the rabbits like best of all, but for which they
are not prepared to make too long a journey.

This picture is rather too simple for some other factors, notably soil, can affect the distribution and composition of the grass heaths of Breckland. The soil may contain much calcium carbonate in places and in consequence be alkaline while in other places it may be acid and deficient in salts. It is not surprising then that some of the grassland is very like that of the chalk since firstly the soil is similar, secondly, both grasslands enjoy a dry climate, and thirdly both are largely maintained by grazing by the same animal, the rabbit. Heather is unable to invade such grassland, nor indeed any soil rich in soluble salts. As the soil becomes poorer grass heath develops with bent and sheep's fescue amongst the principal grasses. In the poorest soils very few flowering plant species are found but lichens have increased to form links between the scattered grass plants.

Breckland is remarkable for a number of plants that are special to it; they are sometimes called 'steppe' plants because their distribution extends to areas of this kind in continental Europe. East Anglia with its low rainfall has a climate more akin to that of continental Europe than any other part of the British Isles. Among the Breckland specialities may be listed the Spanish catchfly (*Silene otites*), the sickle medick (*Medicago falcata*) and some speedwells (*Veronica* spp.), but altogether there are about twenty species with a distribution largely centred on this part of England.

While heaths are characterised by particular plant communities commons are open land of one kind or another on which certain people have common rights and which are, in general, unfenced and available for everybody to walk and explore. They are quite often on poor soils and if these are sandy they tend to become, in effect, heaths. If the area is cut regularly then it is most likely to become a grass heath like parts of Hampstead heath or Wimbledon common, both in or near London. If the soil is clay, then a common may well contain some areas of woodland like Bookham and other Surrey commons and also some damp marshy spots as well as grassland. Areas of common land vary greatly in size for they may range from tiny village greens to areas of thousands of acres. They include forests, woodland, moorland fen and marsh and they altogether amount to about a million and a half acres in England and Wales. Though a reservoir of common wild plants, many hold a special plant of local if not of national interest, and who knows what may not yet be found by the careful searcher with a really keen eye for the unusual?

CHAPTER 10

HEDGEROWS, VERGES AND PATHWAYS

EVEN today, with the great expansion of building that has gone on in this country in and around all cities, towns and villages, it is rarely very far to a footpath, or roadside verge or a hedgerow. It may be only a pathway through a public park or a hedge along the side of a recreation ground but it can provide an introduction to some common plants. Alternatively you may be able to enjoy easy access to some delightful country paths through fields and woodlands or by streams and riversides. You may be even more fortunate and be within easy reach of hills and mountains with their own special vegetation which many have to journey miles to see. But let us start with the kind of hedgerow so common in the lowlands of Great Britain.

A typical hedge and road side often shows several distinct habitats: there may be a narrow strip of meadow along the road, then perhaps a ditch or depression, then a bank on which a hedge itself is growing, and each one of these places has plants of its own. The actual species present depend on many factors; on a pathway you will find only one or two grasses and some other plants that can withstand much treading. Annual meadow grass (*Poa annua*), the commonest grass in Great Britain, is almost certain to be there; it may be recognised by its tufted annual habit, flattened shoots and spreading flower panicles which can be found in any month of the year. One or two plantains with their flat rosettes are also very frequent; one, the ribwort (*Plantago lanceolata*), has narrow strongly veined leaves while the greater plantain (*P. major*) has broader leaves with short but distinct leafstalks. The rosettes of the creeping and meadow buttercups (*Ranunculus* spp.) with their divided leaves may well be seen and there are other species such as the docks that may be seen here and there. The pathway may show a zonation towards the outer edges. In the middle there will almost certainly be a central zone where the treading prevents anything from growing, while towards the edges there will be plants more tolerant of treading and a gradual transition to the surrounding meadow-like verge may be seen.

The plants of a roadside verge can be varied indeed; they may range from the commonest of weeds to rare plants of almost national importance. Their presence will depend partly on the treatment the verge has

FIG. 27. *Left*, the greater plantain (*Plantago major*); *right*, the ribwort (*P. lanceolata*).

received and anyone interested in a particular area should endeavour to discover exactly what that has been. Usually it is cutting and mowing of some kind which prevents the development of shrubby plants. The verge may also be sprayed with weedkiller, a practice that is aesthetically intolerable where all the plants are killed, but perhaps less so when it acts selectively or serves to cut down the rate of growth of the plants. Roadside verges show a wealth of plant growth in spring, and some of the most attractive of the regularly occurring plants include the cow parsley (*Anthriscus sylvestris*) with finely divided leaves, tall stems growing three to four feet or more and masses of white flowers. Altogether it is a very elegant plant and worthy of its other English name of 'Queen Anne's lace'. About the same time a common crucifer garlic mustard (*Alliaria petiolata*) with white flowers and pale green leaves smelling strongly of garlic on being crushed is common enough. There

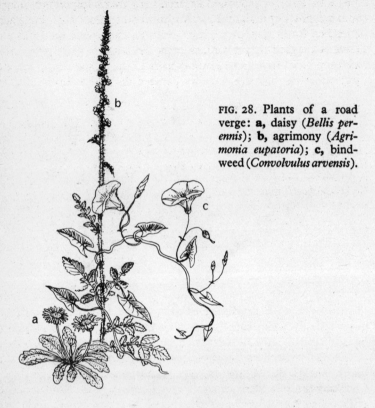

FIG. 28. Plants of a road verge: **a,** daisy (*Bellis perennis*); **b,** agrimony (*Agrimonia eupatoria*); **c,** bindweed (*Convolvulus arvensis*).

will be plants like the speedwells with their bright blue flowers, and later others like agrimony (*Agrimonia eupatoria*) with long spikes of small yellow flowers which form hooked fruits that catch on to clothing. The members of the Compositae are well represented, for they include the dandelion (*Taraxacum officinale*) and daisy (*Bellis perennis*) and others a little harder to distinguish. One of the earlier yellow-flowering composites to look for is the cats-ear (*Hypochaeris radicata*) which has a rosette of rough lance-shaped leaves from which spring a number of almost leafless stems carrying the yellow flower heads. The yellow flower head will be found to have a number of chaffy scales between the florets and this serves to distinguish the few species of *Hypochaeris* from all the other yellow-flowered composites.

Careful examination of the fruiting head of the cats-ear will show the pappus of silky hairs by which the fruit is blown about. Many of the composites have a pappus and its details are always important. Sometimes it consists of straight silky hairs but at other times it has branched or feathery hairs. *Hypochaeris* has a feathery pappus and the fruits are drawn out at their tips into a long point or beak. The hawkbits (*Leontodon* spp.) also have some feathery hairs to their pappuses which are dirty white but there are no scales between the florets. In the thistles

FIG. 29. *Left,* the fruits of a hawkweed (*Hieracium pilosella*) with simple hairs and no scale. *Right,* the fruits of cat's ear (*Hypochaeris radicata*) with feathery pappus and scale.

the two major genera are distinguished by the nature of their pappus, the genus *Carduus* having simple hairs while the genus *Cirsium* has feathery hairs. Common wayside thistles include the welted thistle (*Carduus acanthoides*) as well as the spearthistle, the creeping thistle and the marsh thistle; the last three all belong to the genus *Cirsium*.

On any roadside there are usually piles of road-mending material or grit or sand at various intervals. These are temporary and so get colonised by many species like the creeping cinquefoil (*Potentilla reptans*) or silverweed (*P. anserina*) or the field bindweed (*Convolvulus arvensis*) and many others; in fact their study is worth while of itself. Some less distinguished plants like knotgrass (*Polygonum aviculare*) and goosefoot (*Chenopodium album*) will most likely be seen here.

Moving from the roadside verge towards the hedge proper, we may well see a depression or shallow ditch where according to the depth of water there may be a number of damp-loving plants. If there is only a slight depression nothing more than a more luxuriant growth of the existing plants may be seen. Where there is real dampness there may be plants like the willow herbs with narrow leaves and pink flowers. The commonest is the great hairy willow herb (*Epilobium hirsutum*) growing four feet or more and with flowers half an inch or so across. Accompanying it, if there is a good depth of soil will be plants like meadow

FIG. 30. Two plants of shallow ditches: *left*, willow herb (*Epilobium hirsutum*); *right*, meadow sweet (*Filipendula ulmaria*).

sweet (*Filipendula ulmaria*) with pinnate leaves growing three to four feet high and crowned with white flowers. Marsh thistle (*Cirsium palustre*), tall and dark with clusters of small dark purple flower heads is another likely plant to be seen here together with some of the damp-loving grasses and sedges.

On the further side of the ditch and growing under the shelter of the hedge there may be a sloping bank. This is the kind of place where ferns like the male fern (*Dryopteris filix-mas*) or the broad buckler fern (*Dryopteris austriaca*) may be found. In the male fern the frond is broadly lance shaped in outline and the fronds are twice pinnate whereas in the broad buckler fern the outline is more triangular and the fronds are tripinnate. If the soil is on the whole calcareous this is a very likely spot for the common cuckoo-pint (*Arum maculatum*) with its arrow-headed leaves and very curious hooded flower spikes. Herb Robert (*Geranium robertianum*) and other species of cranesbill are likely to be found here. The two commonest cranesbills are perhaps the cut-leaved (*Geranium dissectum*) and the dove's foot with rounder and less divided leaves. All species have the long pointed fruit from which the plants get their name.

Within the shelter of the hedge itself is the place where climbers and scramblers are to be found. One very common plant of this kind is goosegrass (*Galium aparine*) or cleavers with all its parts covered with

FIG. 31. Two plants of the hedgerow: **a,** black bryony (*Tamus communis*); **b,** cleavers (*Galium aparine*)

tiny hooks which readily catch on to other plants and enable it to climb. The leaves are narrow, six to eight in a ring or whorl, and the stems are four-angled. The flowers are very small, about an eighth of an inch across and greenish white. Other climbers of the hedgerow are the bryonies, black and white, of which the former (*Tamus communis*) is the more frequent. This has heart-shaped leaves and stems that twine round their supports. The ripe berries of this plant are a bright scarlet and beautifully translucent. White bryony (*Bryonia dioica*) climbs by means of tendrils and has lobed leaves; its flowers are greenish white and the berries also turn red when ripe. Both these plants have massive underground rootstocks or tubers though botanically they are unrelated. A climber of another kind is the woody nightshade (*Solanum dulcamara*) and it is known as a scrambler because of its way of climbing. It produces shoots that grow between the branches of the hedgerow and by lodging here and there succeed in reaching the top without possessing any obvious mechanism. Woody nightshade also grows in quite different habitats, for instance, in shingle on the sea coasts or sometimes in quite wet places like the edge of a pond from which it can even spread into the water. All the climbers must firstly, be physiologically adapted to grow rapidly in the centre of the hedge and secondly, adapted to exploit the full light that is to be found at the top of the hedge. Plants of the open would fail to fight their way to the top of the hedge and die back before reaching it.

Many of the shrubs that make up a hedge have been mentioned in Chapter 3 and it will not take long to pick out and name the shrubs of any hedge. If, however, you look at several hedges you will find that some contain more woody species than others. One hedge may be almost pure hawthorn while another may contain hawthorn, sycamore, maple and hazel for example. One can think of reasons for this. If one plants a hedge to-day it is usual, easy, and convenient to plant one species only like thorn, holly or privet. If so any resultant diversity must be the result of subsequent invasion by other shrubby species. Alternatively the hedge could have been made by planting several different kinds of shrub. However, the diversity of shrubs in hedgerows has been thoroughly studied recently by listing the number of woody species in thirty-yard lengths of hedgerow. In order to get a fair figure, more than one thirty-yard sample length was sampled for each hedgerow and some were found to have as many as twelve woody species in a thirty-yard length. In some cases it is possible to date hedges from historical sources such as Anglo-Saxon land charters, parish boundaries, medieval records and estate maps. When these two sets of facts are put together a simple conclusion emerges, namely, that the older the hedge the more species it contains, and that broadly speaking, one species corresponds to about one hundred years. Thus a hedge with three woody species may be nearly 300 years old, while one with as many as ten will be about 1000 years old. There will be some exceptions to this, for no doubt some farmers may well have planted three to four species of hedges in the beginning, but this seems to have been unusual. Dr Hooper of the Nature Conservancy's Monks Wood research station who initiated this work has investigated hundreds of hedges that can be dated with some degree of accuracy and concludes that the correlation is well justified.

What is the reason for this and how has it come about? The answer is to be found in the facts of succession mentioned in Chapter 2 where it is pointed out that, if an area of ground is left, new species tend to arrive and colonise the area and so the composition of its flora changes. Likewise, if a straight row of hawthorn is planted, in due course other woody species will gain a foothold and so the composition of the hedge will alter. It will tend to become rather more like a strip of woodland, but of course the cutting of the hedge itself will prevent the development of trees.

In a plant succession there is often a definite sequence and species follow each other in almost an orderly manner. On a chalky slope one of the first plants to arrive might be holly, yew or hawthorn, later to be followed by ash and still later by the beech. A similar sequence is to be seen in the hedgerow succession. Field maple (*Acer campestre*), for example, is not often found in a hedge that has only one or two species in it, but is found in a hedge with about four species in it, while spindle (*Euonymus europaeus*) is found in hedges with about six species in them. It seems therefore as if the conditions are not right for the establishment of maple until the hedge is about 400 years old and it has to be even older before spindle is able to establish itself.

If land is left to itself it succeeds to some kind of woodland in about a century which is much shorter than the time that a hedge takes to reach its maximum diversity. This is a difficulty but it should be remembered that the hedge is a special man-made affair; it is a very narrow and complete strip and new species can only invade when one of the shrubs making it up dies. This must limit colonisation, and in view of this, the long time taken for the hedge to become diverse does not seem so great a difficulty. Here is a record for two hedges, one at the back of the house where the author lives and another at the front. The hedge at the back has eight species in it namely

ash	hawthorn
dog rose	hazel
elder	sycamore
maple	Norway maple

which would make it 800 years old; the hedge immediately in front of the house is known to have been planted about 100 years ago and it has three species, hawthorn, holly and green privet, while a little further on it has more. A reference to maps shows how the property was originally enclosed from the common land, so the hedge at the back is presumably part of the old common land boundary dating back to Norman times, while that immediately in front of the house can be

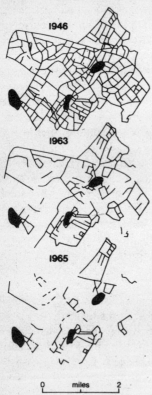

FIG. 32. The loss of hedgerows in three Huntingdonshire parishes. After Hooper.

dated accurately; the part a little further on is more diverse but a little short for accurate counting but it may well date back to one of the old cottages that once stood on the site possibly 300–400 years ago.

The composition and dating of a hedge is therefore of great interest, particularly as it affords a link between botanical and historical studies; it can lead on to more than this, for the making of the English countryside has been influenced in so many ways, economic as well as historical.

There have been changes in the number of hedgerows in the past centuries; a great many were planted as a result of the enclosure acts of the eighteenth and nineteenth centuries. Now the reverse is happening and hedges are disappearing at a rate estimated at something between 3,000 and 4,000 miles each year. Though the rate of disappearance varies in different parts of the country, and it is not easy to estimate accurately, the figure given is about the average of those found in various investigations. Making large fields increases the area of cultivable land by the area of the hedge removed and it saves the cost of maintenance of the original hedge. It is also easier for modern farm machinery to work large fields, for the amount of turning round is reduced and there is a diminution in the loss of material through overlapping in working. Square fields of about 50 acres seem to be near the ideal from this point of view and in the flat cereal-growing land a complete change to this state of affairs has almost been achieved.

The disappearance of hedgerows is unlikely to lead to the extinction of any woody species, but it is as well to realise that common hedgerow plants are biologically important since they are the food plants for many other organisms. To quote an example, the average plant in the British flora can support about six species of moth but the oak can act as food plant for 114, and elm, ash, hazel and rose can support comparable if not quite so large numbers. Numerous hedges, too, provide habitats for many larger animals, in particular birds, and if it is accepted that 100 yards of hedge supports a pair of birds, then this means a loss of something like twenty pairs per mile and 8,000 pairs per year. The effect of this would vary greatly according to the species but it might be expected to affect the total population of whitethroats greatly since they nest in hedges but to have less effect on the nightingale population which is mainly a woodland species.

Furthermore, the really old hedges with ten woody species or more in them date back to Saxon times and are boundaries of great historical interest. There are therefore good reasons for preserving some, at least, of our older hedgerows and this is encouraged by all conservationists.

CHAPTER II

MARSHES AND FENS

MARSHES are usually found where the soil is waterlogged and where the water is moderately rich in dissolved materials. Where there is water standing at some time of the year above the level of the soil what is called a swamp flora develops. The presence of dissolved minerals is one of the reasons for the very great difference in the floral composition of

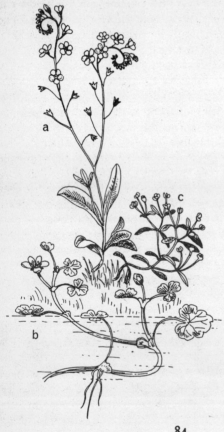

FIG. 33. Three marsh plants: **a,** water forget-me-not (*Myosotis scorpioides*); **b,** water crowfoot (*Ranunculus omiophyllus*); **c,** water blinks (*Montia fontana*).

marsh and bog. The contrast between bog with its bog moss (*Sphagnum* spp.), purple heath grass (*Molinia coerulea*) and cotton sedge (*Eriophorum* spp.) and the marsh with marsh marigold (*Caltha palustris*), forget-me-not (*Myosotis* spp.) and other species is very striking.

Narrow fringes of marsh can often be found round the edges of ponds, along the sides of rivers and in the damp hollows of fields. Where a river has spread widely in flood so as to form a flood plain, quite a wide expanse of marsh may occur with numerous species of plant. There may be the bright yellow buttercup-like flowers in early spring of the marsh marigold which may be distinguished from true buttercups by their relatively large heart-shaped leaves. Actually, too, the structure of the flowers is different for the yellow floral parts are really sepals, and the fruit is a kind of pod that splits down one side. Among plants also found in marshes, one thinks of ragged robin (*Lychnis flos-cuculi*), the cuckoo flower (*Cardamine pratensis*) and the yellow daisy-flowered fleabane (*Pulicaria dysenterica*). Water forget-me-not (*Myosotis scorpioides*) with its bright blue flowers is also characteristic and so are some species of the Labiatae like the water mint (*Mentha aquatica*) known by its strong smell and gipsywort (*Lycopus europaeus*) which is white-flowered with deeply cut leaves. Members of the Umbelliferae are well represented, particularly by the water dropworts, fool's watercress, and others. Chief among the dropworts is the hemlock water dropwort (*Oenanthe crocata*), a stout perennial growing three to four feet high with very divided leaves and numerous small white flowers is flat panicles. The roots of this are spindle-shaped and poisonous and have been mistaken for horseradish with fatal results. Most of the dropworts are plants of wet places where they should be looked for, though the distinction of some of the less common ones is not always so easy. One species, *Oenonthe aquatica*, with very finely divided leaves, often grows partly in water but it can be found on mud by riversides; another, *Oenanthe fistulosa*, has swollen stems rather reminiscent of an onion, while one or two other species should be sought in damp meadows by the sea. Another plant which at first sight does not resemble an umbellifer is marsh pennywort (*Hydrocotyle vulgaris*); this has creeping stems and large numbers of round leaves which make it fairly obvious, but the flower heads are very small and hidden beneath the leaves. This latter plant is more frequent where the soil is a little peaty, and in such places may be found the lesser spearwort (*Ranunculus flammula*), the sneczewort (*Achillea ptarmica*) and others. Sneezewort is related to common yarrow, but it has a smaller number of larger flower heads and rather less divided but narrow leaves. Rushes and sedges will be well represented, for they flourish in waterlogged soils. On basic soils hard rush (*Juncus inflexus*) will be the most frequent, while soft rush (*Juncus effusus*) will be the more abundant where the soil is more acid and peaty. Sedges too will be seen, like *Carex flacca, C. pendula* and occasionally *C. elata* while near the water and in swamp conditions some of the larger sedges like the greater and lesser pond sedges (*C. riparia* and *C. acutiformis*) may be seen.

Plants growing in waterlogged soils are liable to be short of oxygen for respiration of the roots, since oxygen is only slightly soluble in water. Some marsh plants like the gipsywort (*Lycopus europaeus*) and the purple loosestrife (*Lythrum salicaria*) develop a special aerating tissue on the stems and roots. It appears as white spongy swellings and the tissue consists essentially of cells that have become rounded off from one another so as to make diffusion between them possible.

Marshland can often develop into woodland by the growth of willows and alders. The commonest shrubby willow is the grey sallow (*Salix cinerea*). When the bark is peeled off the two-year-old twigs of this species longitudinal ridges are seen, a character that is shared by few of the other species. One is the eared willow (*Salix aurita*) a small bush characteristic of acid and boggy soils. Some other willows are fairly easy to recognise, the osier (*Salix viminalis*) with very narrow leaves and long wand-like twigs that are used for basket making. Purple willow (*Salix purpurea*) is less common, but its upper leaves are nearly if not exactly opposite and it has purple anthers. Also not very common is the almond-leaved willow (*S. triandra*) which is particularly beautiful when in flower; it has a smooth reddish brown and peeling bark, together with dark green leaves, that are usually paler beneath, though quite variable. Two of the larger willows, the crack (*Salix fragilis*) and the white (*S. alba*) are often pollarded, and new trees are obtained by the simple expedient of putting a willow stake into the wet ground when its roots at one end and puts forth shoots at the other.

Alder (*Alnus glutinosa*) is a tree that develops by river and stream sides and it can form pure woods in places, but it avoids the more acid soils. An unusual feature of the young twigs is that the pith is seen to be three-rayed when the twig is cut across. Another unusual feature of the alder is the possession of root nodules often as large as a cricket ball inhabited by a nitrogen fixing organism similar to those occurring in the Leguminosae though its exact identity is still in doubt. An additional source of nitrogen must be of considerable importance to this tree.

The luxuriant growth of marsh plants may lead to the accumulation of plant remains since under water oxygen supply is minimal and therefore decay is very slow. The water is usually more or less alkaline according to the salts it contains, and so layers of peat may be laid down and true fen formed. The plants growing are similar to those of the marsh from which it has been formed and quite unlike those of bog peat. There are large areas of fen in East Anglia, for example, the Norfolk broads, but fens are less frequent in the western parts of the British Isles, though they do occur where water drains off the limestones and there are also quite large areas of fen in the centre of Ireland and in Somerset.

The final state of the fen vegetation is markedly influenced by the treatment it receives. In Norfolk the tendency is for the swamp flora of common reed (*Phragmites communis*), reedmace (*Typha* spp.) and burr-reed (*Sparganium* spp.) to be entered by the saw sedge (*Cladium mariscus*). This plant is tall and vigorous, growing four feet or more, and it ultimately comes to dominate the fen to the exclusion of other plants. Its leaves are long and have unpleasant teeth along both the

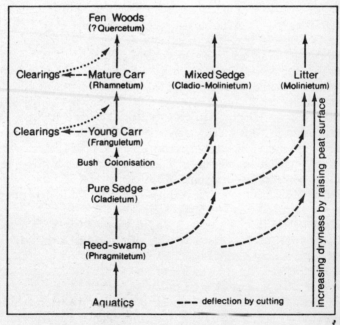

FIG. 34. The plant succession of a Cambridgeshire fen. After Godwin.

edges and the midrib and its spikelets are arranged in dense heads. It was the practice to use this sedge for thatching and other purposes when it was cut every four years; this period of time allows it to recover completely from the previous cutting. If it is cut more frequently, then the sedge is held in check and the smaller sedges, purple heath grass, hemp agrimony (*Eupatorium cannabinum*), angelica (*Angelica sylvestris*) and yellow loosestrife (*Lysimachia vulgaris*) are to be found growing with it. This forms a community known as mixed sedge. When the cutting is more frequent, say once a year, a 'litter' of even more mixed composition is produced and this is used for bedding stock. If the 'litter' is cut more often than once a year, the vegetation comes to be a mixed sward with grasses, small sedges and rushes growing in it.

The saw sedge community if left to itself is rapidly invaded by a number of shrubs like the buckthorn, willows, even guelder rose (*Viburnum opulus*) and others so a thick fen carr is established. This may ultimately proceed to woodland for in certain places alder, birch, ash, and even oak are to be seen developing in carr. So it seems that here we have a complete sequence from open water to woodland.

This sequence does not always take place in nature, even if left to itself, for occasionally something like the development of acid bog takes place particularly if the climate is wet enough (see page 103). There are quite a number of fens in which clumps of bog moss occur here and there even if no further development takes place.

So far nothing has been said about the large open areas of water that have been so marked a feature of the fen country. It is true that many of them were the result of rises in sea level in Roman times as well as falls in the land level. Further, when peat is drained of water it shrinks, and so its level falls. Thus water no longer drains into the rivers but has to be lifted out of the land into the rivers by pumps of one kind or another. Partly as a result of the difficulties of draining large areas of water known as meres persisted for a long time. Whittlesey mere in Huntingdonshire was not drained until the last century. Now only small areas of open water are left and they are disappearing.

It will be seen that the existence of a fenny type of vegetation depends almost entirely on the water level; the vegetation itself tends to raise itself above the water, while man's activities all tend to lower the water level. Therefore to conserve an area of fenland it is essential to maintain a level of water in it which is almost certainly likely to be above that of the surrounding land and this can only be done by the manipulation of sluices and pumps. Wicken Fen in Cambridgeshire is a famous example of this.

The Norfolk broads had a different origin to most fens for they were man made in medieval times by peat digging. Peat digging first began about the twelfth century and continued for over two centuries. It declined gradually owing to the increasing difficulty of getting the peat out of the flooded diggings. Close examination of the shapes of the broads, comparison of maps of various dates, boring and diggings from side to side have all provided evidence that has confirmed these conclusions. In places broken lines of reed swamp marking the position of old parallel baulks that ran across the area can be seen stretching into a broad. Since the cessation of peat digging the broads have been declining in area by gradual silting up and the succession from open fen to carr has been taking place. The increasing use of the Broads for sailing and recreation tends to scour the rivers and broads and helps to keep them open but dredging is necessary in addition to this.

The Broads have a flora of their own. A plant of pondweed type (*Najas marina*) can be seen in soft mud in clear and open water in a few broads and nowhere else in Britain. Hardly anywhere else can aquatic flora be seen in such diversity and abundance. Some of the rare plants like the fen orchid (*Liparis loselii*) still persist, marsh orchids are abundant in places, and the marsh sowthistle (*Sonchus palustris*) thought to be declining, is actually doing more than just holding its own.

CHAPTER 12

WATER PLANTS

PONDS are the most frequent areas of water to be seen in Britain, and most of them have been made by man some time in the past to provide water for his farm animals. To-day, particularly where a piped water supply is available ponds are neglected so they silt up and pass through a marshy stage to dry land. Wayside ponds may also become depositories for rubbish of all sorts of kinds, and they may be near enough to roads for oils and tar to pollute them. Thus for ponds to remain as areas of fresh water supporting plant and animal life maintenance is essential.

Most ponds contain a few water plants and fringing the edge some swamp and marsh plants, though the number and quantity of the latter depends very much on the amount of treading and trampling round the edges. Where this is extensive, the soil may become completely bare and give quite a false impression of what the fringing vegetation might be. The aquatic species to be found in any pond depend partly on the size; the larger the area of water the more species of plant occur, at least up to a point. The number also depends on other factors like the nutrients available and the immediate climate of the pond. Some small ponds in farmyards may be so fouled by cattle that next to nothing will grow in them. Others may be so shaded by surrounding trees that they may be almost without flowering plants. Slow decay means that the bottoms become covered with decaying leaves and other detritus.

It is not surprising that among the most frequently found plants of ponds are some of those like the duckweeds that are most completely adapted to life in water. One line of evolution has led to these small free-floating plants that can easily be carried from one water surface to another by birds. The duckweeds are some of the most reduced flowering plants there are and in fact the uncommon *Wolfia arrhiza* which consists of tiny fronds about one millimetre across is certainly the smallest flowering plant in Britain if not in the world. It has no roots and reproduces by budding off daughter plants and it is not known to to flower in this country. There are four other duckweeds to be found in still waters of this country and they are all easily distinguished. There are three with round fronds of which by far the most abundant is the common duckweed (*Lemna minor*) with flat round fronds and a single root to each frond. Similar to it, but larger, and with several roots to each frond is the great duckweed (*Lemna polyrrhiza*) which has a tendency to develop a purplish tinge on the underside of the frond. Gibbous duckweed (*Lemna gibba*) has a frond that is swollen beneath with a single root to each frond. Ivy duckweed (*Lemna trisulca*) is found below the surface of the water and has several fronds that are thin, oval and succulent and joined by very fine stalks. Duckweeds flower but rarely and then only in shallow water exposed to full sun which presumably indicates that a high temperature is necessary for flower development. The flowers are minute and very reduced and are borne in pockets on the margin of the frond. They are easily overlooked, unless searched for with the aid of a lens.

Another plant common in ponds is the hornwort (*Ceratophyllum demersum*) which is just as remarkable as the duckweeds. It is found

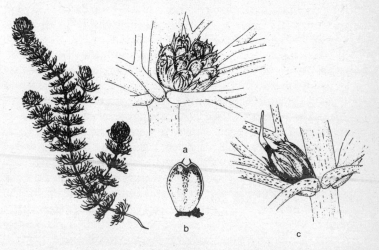

FIG. 35. The floating shoot of hornwort (*Ceratophyllum demersum*) and **a,** the male flower, **b,** a stamen, **c,** a female flower. After Proctor and Yeo.

below the surface, and is free floating for it does not possess any roots and consists of stems bearing forked leaves at the nodes. Since these are rather stiff they have given the plant its name. Its flowers are among the plant's most unusual feature for they are produced in the axils of the leaves and only in the warmest and best-lit waters like those of the duckweeds. The male flowers consist of little more than a group of stamens; the anthers float to the surface where they burst to release the pollen which gradually sinks in the water and falls on the stigmas of the female flowers. This mechanism, completely suited to under water, is made more effective by the habit of the hornwort of growing in masses.

There are other plants with similar pollination mechanisms adapted to life in water. Canadian pondweed (*Elodea canadensis*), a submerged and common water plant easily recognised by its oval leaves about half an inch long that are borne in whorls of three, is just such a plant. The female plant was introduced to Britain about 1840, and this bears single flowers that are carried to the surface on long stalks. The male plant is very rare in this country so the pollination cannot be observed, but the pollen is scattered on the surface of the water and floats to the nearby stigmas. Canadian pondweed, when first introduced spread extremely rapidly and became so abundant that it blocked up the canal systems and caused much trouble to water traffic. Later it settled down to become a normal and common member of the aquatic community, and to-day, though still common, it may be declining slowly. Other plants likely to be fetched out of ponds, lakes and rivers, are the pondweeds (*Potamogeton* spp.) which are delicate, thin and often narrow-leaved plants that dry up very readily when removed from the water, but they can be revived in water for study. One or two species have oval leaves that float on the surface, but in these, the submerged leaves are much narrower. The greenish brown flowers of these plants are borne in cylindrical spikes and in some species they are carried above the water in which case they are wind pollinated, whereas in others, the spikes lie flat near or on the surface and the pollen floats to the stigmas.

There are many other water plants with conspicuous flowers like those of land plants that are carried above the surface of the water to be pollinated by insects. The flowers of the water lilies, the water butter-cups and the water plantains immediately come to mind in this connec-tion. Some of these plants possess leaves that are oval or round that float on the surface of still water and finely divided leaves below the surface. The thin underwater leaves are suited to the absorption of substances like oxygen, carbon dioxide and dissolved minerals in solution and they offer little resistance to moving water. Some plants like the arrowhead (*Sagittaria sagittifolia*) bear three kinds of leaves, narrow strap-shaped leaves below the surface, oval leaves floating on the surface and arrow-shaped leaves right above the surface.

Many water plants possess special means of surviving unfavourable periods. The arrowhead produces purple-coloured tubers from its base which not only survive extremes of temperature but also serve to

FIG. 36. Arrowhead (*Saggittaria sagittifolia*).

reproduce the plant if they are washed from place to place by floods. Other water plants like hornwort and Canadian pondweed readily break into small pieces that will float away and establish new plants. They also produce winter buds that perennate at the bottom of the pond or stream and produce new plants in the following spring. One of the most interesting plants to do this is the frogbit (*Hydrocharis morsus-ranae*) which produces small buds about one third of an inch long which sink to the bottom of the pond. They are weighted at the base so they float vertically and when spring comes they begin to unfold, and their density decreases as respiration uses up the reserves. So they float to the surface to produce a new colony of the plant. A rare plant of East Anglia the water soldier (*Stratiotes aloides*) forms large rosettes of spiky leaves that float near the surface. As the season advances they become encrusted with lime and sink to the bottom of the pond or slow-moving water in which they grow. When spring comes fresh growth occurs and the whole plant is now lighter so that it floats to the surface once more. Water soldier is grown in garden ponds and has escaped in some parts of England, for instance some of the old canal relics in the west of Surrey are full of it. The plant is dioecious,

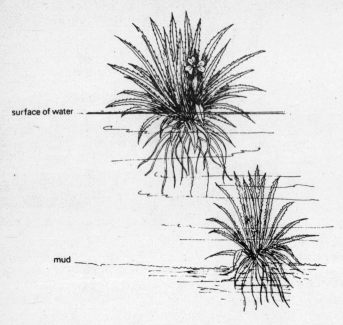

surface of water

mud

FIG. 37. Water soldier (*Stratiotes aloides*) *above*, in summer and *below*, in winter.

and only the female plant is known as native; needless to say it never produces fruit and the plant must spread only by vegetative means.

Large areas of water such as lakes where there is some accumulation of silt may show a wide zone of reed swamp round the edges. This includes many plants rooted in the mud that have excellent means of spreading vegetatively and tall stems that rise above the surface of the water-bearing sword-like leaves and aerial flowers. These include well known plants like the bulrushes (more correctly reedmace) the true bulrush (*Scirpus lacustris*) the yellow flag (*Iris pseudacorus*) the burr-reeds (*Sparganium* spp.) and the common reed (*Phragmites communis*). Between these there may be smaller plants like the marsh bed-straw (*Galium palustre*) growing beneath the protection of the taller species, while in the open water there may be plants like the yellow water lily (*Nuphar luteum*), the water milfoils (*Myriophyllum* spp.) and the pondweeds already mentioned. On the landward side, the plants are usually not so tall, and the region is described as a marsh as opposed to

FIG. 38. The plants of different water levels. *Left to right:* marsh marigold (*Caltha palustris*); flag (*Iris pseudacorus*); common reed (*Phragmites communis*); great reedmace (*Typha latifolia*); duckweed (*Lemna minor*); yellow water lily (*Nuphar lutea*); hornwort (*Ceratophyllum demersum*); amphibious bistort (*Polygonum amphibium*); water crowfoot (*Ranunculus* sp.); arrowhead (*Sagittaria sagittifolia*); bur-reed (*Sparganium erectum*).

a reed swamp. The marsh may include some plants like the rushes, the flote grasses, water forget-me-not and others.

The lakes and tarns in the areas of the north and west of Britain where the water drains off hard rocks and so contains very little soluble mineral matter are quite different. In the Lake District even the lakes with the most silt have only about one seventh of the soluble mineral matter to be found, for example, in some of the lakes of the Norfolk broads.

The lakes of the Lake District show stony bottoms and one of the first colonisers of the shallow parts near the edge is the quillwort (*Isoetes lacustris*) which is a non-flowering plant related to the ferns, and which bears rosettes of long and very narrow leaves. In the deeper parts there are other non-flowering plants like the stoneworts (*Chara* spp.), delicate much-branched plants often encrusted with lime, together with some species of pond weeds and possibly the water milfoils. On the more gravelly bottoms shoreweed (*Littorella uniflora*) is to be seen; this plant produces rosettes in shallow water and since they grow very close together a large patch has something of the appearance of a turf. Growing with it one often sees the water lobelia (*Lobelia dortmanna*)

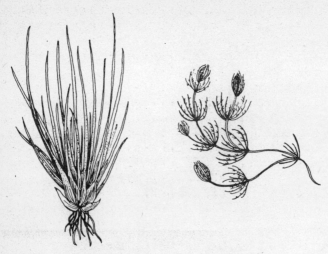

FIG. 39. Two plants of stony lakes: *left,* quillwort (*Isoetes lacustris*); *right,* stonewort (*Chara* spp.).

which has pale lilac flowers borne in spikes above the surface of the water. The fringing vegetation usually shows some transitional stages to bog, for this, rather than marsh, is the more usual form the vegetation takes. As the lake gets older and accumulates silt and more dissolved matter, the number of species may increase with the appearance of some sedges, possibly the bogbean (*Menyanthes trifoliata*) and there may be some of the other plants characteristic of the lowland lakes and ponds.

Rivers, of course, differ from ponds and lakes by their rate of flow of water; if the rate is very slow there is not all that difference between the vegetation of pond and river but if it is very fast there is a world of difference, for flowering plants cannot gain a foothold in mountain torrents, or even in streams that are subject to much flooding, for this will wash out the plants by their roots. There can be everything between these extremes even in one river, for it may start as a mountain torrent and end as a placid stream meandering through a lowland plain to the sea. In the faster moving streams, practically the only flowering plant to be found is one of the water crowfoots (*Ranunculus fluitans*) and underwater forms of one or two plants like fool's water cress (*Apium nodiflorum*). As the rate of flow slows down, pondweeds will be found and nearer the banks the beginnings of reed swamp may begin to appear. For water lilies and similar plants the flow has to be really slow, like one or two miles per hour, till finally a pond-like stage is reached.

Rivers, ponds and lakes are very subject to pollution and indeed have been so since man's early history. Pollution possibly reached a maximum in this country during the Industrial Revolution and subsequently, but since then there has been a slow improvement. The most frequent pollutant of rivers is sewage either treated or untreated, but rivers purify themselves if there is sufficient oxygen supply. As a rule if the water is moving and the concentrations involved are not high, the effect may not be too pronounced. Rivers also gain soluble minerals from agricultural land as the result of the use of fertilisers. Detergents are a further cause of pollution. When the latter were first introduced they consisted of compounds that were not readily decomposed by bacteria and in consequence their presence caused the most unpleasant foaming on rivers. Now most detergents are 'soft' and these are decomposed by bacteria so there is little foaming, but phosphates are added to the water as a product of their breakdown.

As a result of these additions, a lake or pond may change from being poor in nutrients to being rich and this change is known as eutrophication. It results in an increase in the small microscopic forms of life known as plankton since their growth in any water is encouraged by the addition of nitrate or phosphate. The growth of plankton may become so dense as to cut off the light supply to the submerged plants below and if this happens they die. Subsequent bacterial activity stimulated by the abnormal quantity of dead material may use up all the available oxygen causing death of the fish and other water-living animals. It is perhaps important to realise that these effects are produced by adding what are essentially plant nutrients to water. There are other forms of water pollution like the discharge of poisonous substances such as cyanides and the salts of heavy metals that can kill nearly all the plants and animals directly. This can only be stopped by the framing and enforcement of stringent legislation.

CHAPTER 13

MOORS AND BOGS

THE distinction between heath and heather moor is a little difficult to make for one often grades into the other. Heaths are usually found in the lowlands on soils where there is a shallow peat layer about one inch or so in thickness. Heather moors are usually found in upland areas where there is more rainfall and where there is a greater depth of peat ranging from several inches to many feet. Examples include parts of Exmoor and Dartmoor in the south west, the moors of central Wales, and very large areas indeed in the north and west of Scotland. All this vegetation grows on soils that are acid and deficient in soluble minerals, land that on the whole is too poor to be cultivated and that in consequence has been given over to deer forests, grouse moors and the roughest of grazing.

The chief plant that dominates this kind of area is of course heather (*Calluna vulgaris*) of which something has been said already, but on moors it is often accompanied by a number of other shrubs and among them is the bilberry (*Vaccinium myrtillus*). This is a deciduous shrub with green angled twigs growing one to two feet high and bearing oval bright green leaves. The flowers are solitary, of a greenish pink and give rise to a black berry that is covered with a bluish bloom. The fruits are pleasant to eat and are known as blaeberries or whortleberries in Scotland and more rarely as huckleberries. This shrub is rather more tolerant of shade than heather and so it grows under trees forming a lower storey in some oakwoods on acid soils. It is not confined to the north although more common there, but it occurs in some woodlands in in the south and west of England. But above all it is characteristic of rocky sites about 2–3000 ft up where it can almost replace heather. Such sites are often spoken of as bilberry edges, bilberry moors and bilberry summits.

Other plants that accompany heather and bilberry in these situations include some closely related species, for example the cowberry (*Vaccinium vitis-idaea*) which is an evergreen shrub a foot or so high with more or less erect stems. The leaves are oval, dark green above and dotted with glands on the underside. The flowers are borne in clusters of about four and are white with a touch of pink. The fruit is red but is not so esteemed as the bilberry.

FIG. 40 Bilberry (*Vaccinium myrtillus*).

Bog whortleberry (*Vaccinium uliginosum*) is deciduous like the bilberry but the twigs are brownish and round as opposed to the green and ridged twigs of the bilberry. Also while the twigs of the bilberry tend to be erect those of the bog whortleberry are more spreading, even procumbent. The leaves are oval, often a bluish green with a well-marked venation. The flowers are pinkish and borne in clusters of one to four in the leaf axils. The fruit is black but covered with a bloom like that of a plum. Bog whortleberry is a plant of higher altitudes than the bilberry.

The two bearberries are both dwarf prostrate shrubs that sometimes form mats. Bearberry (*Arctostaphylos uva-ursi*) is evergreen with ovate leaves and with five to twelve flowers in rather dense clusters. The fruit is small, red and glossy with a rather dry flesh. The shrub is quite common in parts of the Scottish Highlands. The black bearberry (*Arctous alpinus*) is a much rarer species that is to be found only on the summits of the mountains in the north of Scotland.

Crowberries are not in the same family as the shrubs already mentioned, but they are low evergreen heath-like shrubs with alternate leaves. The flowers are very small, pinkish and the fruits small and black. Cloudberry (*Rubus chamaemorus*) is a herb sending up annual stems with solitary white flowers. It is closely related to the blackberry (*Rubus fruticosus*) and the raspberry (*Rubus idaeus*) but its fruits are orange when ripe and very pleasant to the taste. It is a moorland plant, sometimes abundant in places, as in parts of N. Wales.

Moorland always supports a number of grasses and sedges. Bent grass, mat grass and purple heath grass have already been referred to, but specially characteristic of the damper parts of moorland are the cotton sedges usually but less correctly called cotton grasses. They are familiar and very easily recognised plants because of the numerous white bristles in the flower that elongate after flowering and give the characteristic white tufts or hairs. Two species are particularly abundant, one has a solitary tuft of white hairs and is known as hare's tail

FIG. 41. Cotton sedge (*Eriophorum angustifolium*).

(*Eriophorum vaginatum*), the other has several tufts on a single stem (*Eriophorum angustifolium*) and a creeping habit of growth. The other two rarer species of cotton sedge have a tufted habit of growth.

A related species known as deer grass (*Trichophorum caespitosum*) is really another sedge and it is a densely tufted plant with a yellowish or light brown appearance that picks it out from the darker green vegetation that surrounds it. It is found on bogs, moors and acid peaty places throughout the British Isles.

Bilberry and heather moor generally occupies the drier regions of hills and mountains. Most of the mountain country was covered by forest below 2000 feet before being cleared by man, and this forest most likely contained the bilberry as an under storey and the present bilberry areas are most likely derived from it. In the woodland the bilberry might well have been accompanied by a number of upland plants and some of these might also have persisted after the forest had gone. Plants of this group might include the wintergreens (*Pyrola* spp.), the dwarf twayblade (*Listera cordata*) and a few other rarities confined to the more northern parts of the British Isles.

Upland moors are often maintained for grouse, and in order to keep the heather in a state to feed these birds, it is burnt at intervals of seven to fifteen years. After burning the heather regenerates from the roots and in addition large numbers of seedlings may develop. This practice of burning limits the number of species for only some can survive this treatment and also it tends to favour heather at the expense of bilberry. In another way the number of species is restricted, for when heather becomes old, spaces develop between and under the older branches where different plants can grow. If the heather is burnt then these spaces do not develop. Sometimes the damper parts of moors are drained to encourage the development of heather and if this is not too severely grazed it increases the area further. If these areas are burnt, some of the sedges, and grasses can survive and grow with the regenerating heather. Sheep, grazing on such vegetation, prefer the heather to the coarse and unpalatable grasses, rushes and sedges and so it is possible for the balance to be turned against the heather, and then coarse grassland develops. Nonetheless heather moor covers a very large part of upland Britain where by regular burning the young shoots provide food for the grouse.

In the highlands of Scotland much larger areas are maintained for red deer stalking. These animals graze on almost anything vegetable and do much damage to crops and they also suffer in hard winters from lack of food. Plantations have to be fenced against them and there is no regeneration in the small remaining areas of woodland if these animals are allowed to bite the buds off the young trees.

Much the same kind of story can be gathered from investigations of moorlands other than those in northern England and Scotland. On

Dartmoor and Exmoor there are large areas covered with heather mixed with other herbs, shrubs and grasses. Many of the plants associated with the northern moors do not occur in the south of England, but there are some that are special to the south west. Fine bent (*Agrostis setacea*) with stiff glaucous bristle-like leaves is a grass peculiar to and often abundant in these areas. Heather moor is burnt (swaled is the local word) at just about ten-yearly intervals in order to favour the growth of young shoots for grazing animals. Too much burning results in the invasion of heather by bracken and leads to a poor type of grassland that does not give good grazing. Gorse and bilberry are also diminished by burning and grazing.

As moors become wetter the vegetation passes into bogland and this kind of vegetation is best developed in areas of high rainfall like the north and west of the British Isles. For many reasons the most import-ant plant here is bog moss (*Sphagnum* spp.) of which there are several

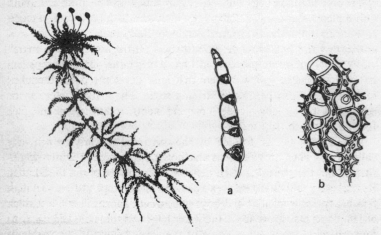

FIG. 42. *Left,* bog moss (*Sphagnum* sp.). *Right,* the leaf structure of a bog moss showing (**a**) section and (**b**) surface view as seen with the aid of a microscope.

species. This plant by virtue of its peculiar structure has an immense water-holding capacity. The leaves are made of long large cells that simply hold water while between them run long thin cells that contain the green chlorophyll and which carry out photosynthesis.

Three kinds of bog can be recognised, namely valley bog, raised bog

and blanket bog. Valley bogs arise in dips and depressions in heath and
moorland into which water can drain and not get away too easily. Little
areas of bogland formed in this way are common in some parts of Britain
like the New Forest, Exmoor and Dartmoor, but rare in the East and
the Midlands. Larger areas occur in wetter climates particularly where
there are any obstructions to the flow of water. Valley bogs contain a
number of very interesting and attractive plants, for example, the sun-
dews (*Drosera* spp.) with their fly trapping leaves. There are three
species; the commonest is *Drosera rotundifolia* with round leaves that
narrow abruptly at the base to form a basal rosette. The long stalked
glands (tentacles) with their glistening appearance make these plants
unmistakable; the flowers are small, white, and borne on an erect stem
from the centre of the rosette. The long-leaved sundew (*Drosera inter-
media*) is very like it but differs in the leaf shape, while *Drosera anglica*
is like the latter but larger in all its parts and with a flower stem about
twice as long as the leaves. The last is the rarer of the three species but
all are only found in bogs where they may sometimes be quite abundant.
Other plants that grow in valley bogs include a St John's wort (*Hypericum elodes*), the heath spotted orchid (*Dactylorchis maculata* ssp.
ericetorum) the bog asphodel (*Narthecium ossifragum*) and the marsh
violet (*Viola palustris*). Bog asphodel has leaves rather like a tiny iris and
it bears a spike of yellow flowers that later develop reddish capsules;
in a mass they can be quite a striking sight. The marsh violet can be
recognised by its long slender creeping stems, round leaves and pale
lilac flowers with dark veins.

Raised bogs can be formed by the continual upward growth of a
valley bog; when this happens the bog area becomes slightly dome-
shaped and the stream is often deflected to one side of the raised area.
Remarkably, too, raised bogs can develop on the tops of fens. Though
fens are typically alkaline and bogs acid, what happens is that a tussock
of fen plants may grow above the water level and there die and decay to
form an acid humus out of reach of the water below. It is possible, if
all conditions are favourable for this acid humus to be colonised by
Sphagnum and other bog plants so that ultimately a raised bog develops. In such cases the remains of the fen plants can be identified in
the peat below and by means of bores made through it a complete
vegetational sequence can often be demonstrated over a very long time.
Raised bogs are most frequently to be seen in Ireland where they are
exploited for their peat but they occur in parts of Wales, Shropshire
and other northern and western areas.

Cranberry (*Vaccinium oxycoccus*) is a typical shrub of raised bog

where the stems creep over the surface, bearing small oval to oblong
evergreen leaves. The pink flowers are in small clusters and the red
fruits are very rightly esteemed for their flavour. Andromeda (*Andromeda polifolia*), a related plant, is another creeping evergreen shrub; it
has nodding pink flowers about a third of an inch across, but the fruit
is a capsule and not fleshy. As the level of the bog goes on rising it
tends to become drier on its upper surface when it may become colonised by heather and ultimately by trees. Blanket bog is found where the
climate is very wet throughout the year, and where there is little
drainage, so the whole is permanently wet. The name blanket bog is
very apt, for in western Ireland, west Scotland and parts of England
like Dartmoor, it literally appears to cover the landscape like a blanket.
It is rather more variable in composition so far as the plants are con-
concerned, for there may be parts covered with heather, or cotton sedge,
or deer sedge, and the purple heath grass is always present. In certain
favoured spots there may be many of the species mentioned earlier as
being found in valley and raised bogs.

The blanket bogs that cover the upper parts of Dartmoor contain a
great deal of purple heath grass as well as deer sedge, cotton sedge and
heath grass and the rest. It seems that the relative abundance of these
plants is due to the practice of burning the moor; Dartmoor is dry
enough at certain times for this to be done and the plants mentioned
survive this treatment better than others, for the new buds by which
new growth occurs are protected within the tussocks of these plants.
This together with the mixed grazing by sheep, cattle and ponies seems
to be one of the main causes of the mosaic of vegetation to be seen on
Dartmoor.

The great blanket bogs of the west of Ireland have their special and
beautiful flora and are well worth making a journey to see. They include
St Dabeoc's heath (*Daboecia cantabrica*) with large rose-coloured
flowers, and one or two other species of heath not known elsewhere
in Great Britain. In the pools there is the pipewort (*Eriocaulon
septangulare*) with globular heads of flowers which is a monocotyledon
though it looks a little like a composite at first glance, and with it there
will be the water lobelia (*Lobelia dortmanna*) which is always a pleasant
plant to see. But apart from blanket bog, the whole area is full of sur-
passing interest to the botanist with a number of unusual habitats and
special plants to see.

CHAPTER 14

WEEDS AND WASTE GROUND

THERE are many, in fact, far too many places where a man or men have
been working and have left waste land behind them that they could not
be forced to tidy up. Such places range from a small heap of rubbish to
a mountain of waste from a coal mine, or from a small hole by the wayside to a great pit many acres in extent. Anyone familiar with the English
scene can recall to mind slag heaps in the mining districts, huge clay
and gravel pits in the lowland areas, great mountains of china clay
residues in parts of Cornwall and large areas of industrial dereliction
in South Wales. Some of these heaps are sources of substances that
pollute streams, rivers and land while others can be more directly
dangerous to man as at Aberfan, while all are a national inherited
disgrace.

Nature, however, tends to repair the damage done by man, and to
clothe all such areas with a growth of plants. Some such places where
soil has just been thrown away may be comparatively fertile and
capable of supporting plants making relatively high nutritional demands.
Furthermore, if the surface is bare, there will be little competition and
so many different species of plant may colonise such areas. Plants that
are deposited with the rubbish itself either as rootstocks or seeds clearly
have an advantage and this is why clumps of the garden iris and the
michaelmas daisy are seen so frequently on village tips and town refuse.

Where there is bare ground the race is to the swift and the plants that
get there first can exploit the absence of competition. Not surprisingly
such plants often have excellent means of dispersal. These include
members of the Compositae with their plumed fruits. One can think
of the common groundsel (*Senecio vulgaris*) and the dandelion (*Taraxacum officinale*) both too well known to need any description. Thistles are
common enough in such places, and the common spear thistle growing
three to four feet with very spiny leaves is one of them; look for the
long segment at the end of each leaf that is characteristic. The spear
thistle is a stout tap-rooted plant usually growing by itself, and is not
to be confused with another equally common thistle the creeping thistle
(*Cirsium arvense*) which by virtue of its many underground creeping
roots is always found in clumps; it has rather smaller pale purple heads

grouped in a panicle while the heads of the spear thistle are a reddish purple. Other Compositae of waste ground are the daisy-like chamomiles and mayweeds with their finely divided leaves. They are difficult for the beginner to distinguish, but the commonest is the scentless mayweed (*Tripleurospermum maritimum*). To be certain of these plants the reader is referred to books more strictly concerned with identification than this, but the first thing to determine is whether there are fine scales between the florets in the head. These are missing in *Tripleurospermum*, *Matricaria* and *Chrysanthemum* but present in *Anthemis*. *Tripleurospermum maritimum* is scentless and this is a further aid to identification. A common *Matricaria* of waste ground is *Matricaria matricarioides* or pineapple weed recognised by its scent and by the absence of any white ray florets.

Docks are often plants of waste ground and although they are fairly

FIG. 43. Plants of waste ground; *left*, curled dock (*Rumex crispus*); *right*, shepherd's purse (*Capsella bursa-pastoris*).

easily recognised by their erect habit, their panicles of small greenish or brownish flowers, few people bother to distinguish the various species. The two commonest species of waste ground are the curled dock (*Rumex crispus*), with leaves narrowed at both ends with crisped or curled edges, and the broad-leaved dock (*Rumex obtusifolius*) with broader leaves that are heart-shaped at the base. If the flowers are examined closely the most conspicuous parts will be seen to be three segments (perhaps corresponding to petals) that are sometimes very toothed along the edges, and which may or may not bear little swellings or tubercles. In the curled dock the margins are not toothed and each usually bears a tubercle whereas in the broad-leaved dock the margins show long teeth and only one has a tubercle. These two docks are often serious agricultural weeds, partly because of the ability of their roots to form buds, so that any piece left in the ground can grow. Other species of dock are less common but they may well occur on waste ground.

Other very common weeds of waste ground include species of *Chenopodium* or goosefoot. These have panicles of greenish flowers, but these are made of five, not three small parts and in general the plants tend to have a slightly mealy appearance. Much the commonest is common goosefoot or fat hen (*Chenopodium album*) with lance-shaped leaves and a rather narrow habit of growth. Other species are similar, but differ in the leaf shape, colour and form of the stem and again reference to more detailed descriptions in other books is desirable to know them well.

Though brick walls, bridges and the like are not to be regarded as waste places, they have an interesting flora; most of the species that occur have a good dispersal mechanism (e.g. the Compositae) and an ability to survive drought. Some ferns, e.g. rusty back (*Ceterach officinarum*) are to be looked for on the damp and shaded sides of walls.

It must be admitted that the number of plants that can be found on waste ground is a very large one and any botanist wishing to know them is facing no small challenge to his powers. The greatest complications come from the presence of many alien species that can take advantage of the open habitats to flourish for a short time, but which never gain a real foothold in the British flora. Ultimately as more plants arrive, the ground becomes completely covered with vegetation, and the competition greatly increases and fewer species survive. In fact as succession takes place herbs may give way to shrubs and even woodland in time.

It is often very difficult to say whence the alien species have come. There is a great number that are straightforward garden escapes. Another very well-marked group comes from bird seeds. One of the most common grasses introduced in this way is Canary grass (*Phalaris canariensis*) known by its oval heads that are whitish in colour and marked by green veins. Rough dog's tail grass (*Cyonosurus echinatus*) is another such rarer plant with a one-sided type of inflorescence, the members of which show long hair points (awns). One way, at least, of getting to know some of the species that may be broadcast by birds is to buy some packets of different kinds of bird seed and sow them, and then try to identify the plants that come up.

Impurities in seed supplies are always a source of alien plants. Vast quantities of grain, for example, are imported for feeding poultry as well as for agriculture, and pure as the seed may be, the quantities are so great that quite a number of individual plants may reach the country in this way. Thus samples of the grass seed of rye grass (*Lolium perenne*) may contain as much as 2.8% of weed seed by weight, which means the number of weed seeds sown with this grass is about six per square yard. When a grain with a similar order of impurity is used for feeding poultry, it will be seen that here there is a potential source of thousands of aliens, if any of the residues ever find their way to a rubbish tip. Even if imported seeds do not find their way directly to a rubbish heap, they may do so at a second remove, so to speak, by growing in a crop in the first place and then spreading to a heap after that.

Another source of aliens is wool waste or shoddy that is widely used as a slow-acting manure by market gardeners. Wool is subjected to quite a drastic treatment in the course of preparation but despite this, the residues still contain seeds that will grow. They come from seeds and fruits that have been caught in the animal's coats in the first place. Such 'wool aliens' quite frequently turn up on large tips of town and county refuse and include plants like medicks and clovers with hooked fruits and teeth and the cockleburs (*Xanthium* spp.) with large hooked fruits. A few of these have become naturalised and form part of our flora. These include the piri-piri (*Acaena nova-zelandiae*), a member of the Rosaceae, which has a round head of fruits each of which has four barbed spines by which they are dispersed. It is a low much-branched creeping undershrub and it is to be found in parts of Kent, Norfolk and on Holy Island off Northumberland.

Apart from tips of household refuse near towns and villages there are other types of waste ground. In some ways railway tracks and embankments are examples, for the development of many shrubs is prevented and so some of the ground is kept open for colonisation by annuals and other weeds. Where railways have been abandoned and the tracks taken up, there immediately follows a great variety of species depending on the nature of the ground and of the surrounding vegetation. Railway

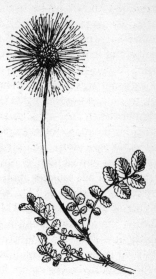

FIG. 44. Piri-piri (*Acaena nova-zelan-diae*, formerly *A. anserinifolia*).

side grassland and scrub was controlled by burning and also by hand cutting. When a railway cutting was made, the subsoil was exposed and when a new embankment was constructed, soil from nearby cuttings was used or alternatively soil was brought from elsewhere. So the combination of these factors created varied habitats for the growth of plants.

Plant development on the permanent way is often prevented by weedkillers but, when the track goes out of use, the top layers of coarse chippings are removed leaving a cinder ballast behind. This is colonised by a variety of plants from a distance as well as being invaded by plants from the nearby grassland. Some of the colonists may come from materials carried by the railways themselves and a good example is the spread of wool aliens by the transport of shoddy by rail. Other aliens found growing in the vicinity of railways may have come with imports of grain and timber.

The tips left by coal mining may be quite small, especially the older ones, but newer tips may be as much as half a mile long and even 200 ft high. In the Derbyshire coalfield these tips are largely shale and coal waste and they readily get colonised by plants with wind-borne seeds (e.g. rosebay willow herb (*Chamenerion angustifolium*)), birches and willows, grasses like Yorkshire fog (*Holcus lanatus*) and some thistles and docks. But the flora can be quite varied and many an interesting find may reward the patient searcher.

Sometimes the soils of heaps seems to favour a few uncommon species of plant particularly well. The mountain sandwort (*Minuartia verna*) is found most frequently on spoil heaps from old lead mines, and the same is true of alpine pennycress (*Thlaspi alpestre*). By contrast there are areas of industrial dereliction in parts of Great Britain where the soils are so poisoned by the residues of past smelting and other industrial processes that hardly anything will grow. Such an area is the Lower Swansea valley in South Wales which was once pleasantly wooded. The timber was first used for coal smelting; to this activity was added copper smelting and later zinc smelting. The smoke from these processes killed all the vegetation and the accumulation of residues of heavy metals and sulphides effectively poisoned the soil.

Today, efforts are being made to rehabilitate such areas. It is possible to landscape these areas with modern machinery and to improve the very low fertility by the addition of some fertiliser but this is hardly enough as organic manures are also required. Various combinations of sewage sludge and domestic refuse have been tried with some success, but any process has not only to be successful but cheap. It is virtually impossible to remove the heavy metals, and the best answer seems to be to find species or varieties of plants that have a wide tolerance of these poisons. So areas around old mine workings have been searched for varieties of grasses in the hope of finding such tolerant varieties. This search has been reasonably successful, and amongst species of grass such as creeping bent (*Agrostis stolonifera*) and brown top (*Agrostis tenuis*) varieties have been found that will stand up to heavy metals in the soil. Experiments have also been carried out to see if any trees will grow on these soils and Japanese larch, Corsican pine, silver birch and alder are amongst those that have tried with some success.

Similar efforts have been made to rehabilitate pit heaps in other parts of the country. Where the heap is old and the material has weathered sufficiently, tree planting is usually successful and the landscape can be transformed. Pines give some green colour in the winter but they can be gloomy and monotonous when planted in large masses. The broad-leaved trees encourage the growth of grasses and other plants below them in a way that evergreens do not, and the greater variation in their appearance through the seasons make them more attractive. Thus the silver bark of birch trees brightens the area in winter and offsets the dark colour of the pit heaps themselves.

Raw shales and similar materials cannot be afforested at once; they must either be improved by the addition of soil or left to weather for some time. After this it seems that lodgepole and Corsican pines are among the best conifers to plant while grey alder and birches are the easiest of the hardwoods to establish. This rehabilitation and afforestation is very important, particularly in industrial Britain, so that the quality of the landscape does not deteriorate but rather steadily improves, and so that these regions become better places in which to live, work and play.

CHAPTER 15

MOUNTAIN VEGETATION

ARCTIC alpine vegetation is almost entirely confined to the tops of high mountains in Great Britain, and, as the name suggests, it consists of plants that grow either in arctic or alpine regions or both, as well as occurring in Britain. Some of these plants have features that make them particularly attractive to gardeners. They include, for one thing, a large percentage of low-growing shrubs, and a number of plants with particularly neat growth habits that form rosettes, cushions or matted growth near the ground surface. When large and attractive flowers or flower heads are added to the neat growth, we have plants that are fascinating to botanist and gardener alike, and that have led to the formation of alpine gardening and other societies devoted especially to their study.

Arctic alpine vegetation in Great Britain has to be sought for; an effort is needed to climb the mountains to see it at its best, and moreover, since the majority of people live in lowland Britain, they have to make a considerable journey in the first case to reach the mountainous districts before they can explore them. Most of southern Britain is under 1000 ft in height; Leith Hill in Surrey, with the aid of a tower on the summit, just reaches 1000 ft, but this height is not enough to make any difference to the vegetation at the summit. It is just the same as it is at the bottom, except perhaps for ferns which tend to grow on the ground at height, but on sheltered walls at the base of the hills. Dartmoor in the south west reaches 2000 ft, but even this height does not allow the growth of any alpines. In fact, as we have seen, the height, together with the heavier rainfall and great humidity, make for the development of blanket bog over much of the summits of Dartmoor. It is only when the mountains exceed 3000 or 4000 ft that the arctic alpine vegetation begins to appear as in parts of Snowdonia, the Lake District and the mountains of Scotland.

In earlier times most of Great Britain, including the mountain slopes, was covered with forest and there are still a few places where relics of this ancient forest can be seen. There are one or two small oakwoods like Keskadale in the Lake District, and in the Cairngorms there are still comparatively large areas of native pine forest. But the pine trees thin out as the height of 2000 ft is reached and the forest gives way to a belt of juniper scrub before grading into the arctic alpine areas.

Where the forest has been cleared, the slopes may support rough grassland and sheep, or the land may carry heather in order to encourage the grouse. Plainly the kind of vegetation present depends on the soil, the climate and the treatment to which man has subjected the land. In the wettest parts there may well be bog, too intractable to reclaim; elsewhere the land may be so covered with large stones or rocks that man can make little or no use of it.

The highest areas where the arctic alpine vegetation occurs provide many different 'niches' where plants can grow. Hollows in the mountain side, or Scots corries as they are usually called give protection from the full force of the high winds and these are the areas to search for the arctic alpine plants, for few can grow on the flat surface of rocks and boulders exposed to all the weather. It is where the rocks have weathered and pockets of soil have been formed that the plants are to be found. Some rocks weather a great deal more readily than others forming a soil comparatively rich in minerals, while others like slates wear away slowly and form poor acidic soils. The distinction between these two extremes can quite clearly be seen in Scotland. Ben Lawers, the most famous botanical mountain of Perthshire is an example of the first while the mountains of the Cairngorms are mainly granitic and carry poor acidic soils except on some precipices.

The severe effects of wind and low temperature at high altitudes are to some extent mitigated by the presence of snow. In very few places in Scotland is there any permanent snow but it does fall as early as September to October and it may last until June or July on the highest mountains. Snow is a very poor conductor of heat so that plants beneath it are protected from very severe frost and cold and are usually at a temperature near freezing point. The low growth habits previously mentioned also give the plants protection for they can also obtain some shelter from small irregularities in the ground level.

The prolonged persistence of snow and cold means that the arctic alpine plants have only a very short growing season; it can never last more than three to four months at the most, and so plants must be suited to make the most of this period. By way of compensation, however, slopes can warm up very quickly in the sunshine; in fact there may be very big changes in the temperature and in the humidity over a period of twenty-four hours as anyone who has walked mountains knows. One can easily get uncomfortably hot and yet, within almost no time, the sky may cloud and a sudden storm and wind produce a most chilling drop in the temperature. Mosses, club mosses and lichens are better suited to such abrupt changes and hence are abundant in mountain regions.

One habit shown in a number of arctic alpine plants is called 'vivipary' whereby the flowers are replaced by tiny buds that ultimately drop off and grow into new plants. Grasses like the alpine fescue (*Festuca vivipara*) and alpine meadow grass (*Poa alpina*) show the habit and the inflorescence carries small tufts of tiny green plants instead of flowers. The habit is considered to be an adaptation to the short growing period, it not being long enough to allow for the setting of seed.

Turning more particularly to the flowering plants of the richer soils there are many worthy of mention. One of the most beautiful of the dwarf shrubs is the mountain avens (*Dryas octopetala*) bearing white flowers each of which has eight petals. Its dark green oblong stalked leaves with their toothed margins are equally characteristic. Other very dwarf shrubs include some willows that grow so low that they form a turf on which you find yourself walking. In fact, it is not until you look closely, that you see those plants really are tiny shrubs and perhaps not until you see the tiny catkins that you recognise them as willows. The The two least rare are *Salix herbacea* with very short branches and round leaves that are bright green and shining, and the net-veined willow (*Salix reticulata*) which has darker green leaves with a very marked venation. Both these shrubs form woody mats but there are other mountain willows that are rather taller plants growing one to three feet or more high. Two have very woolly leaves (*Salix lanata*) the woolly willow and the downy willow (*Salix lapponum*) while the myrtle-leaved willow (*Salix myrsinites*) has bright green shining leaves. These taller shrubs usually grow on mountain ledges where the mountain behind gives them a measure of support against the wind.

FIG. 45. A cushion plant of mountains, moss campion (*Silene acaulis*).

There are cushion plants like the early moss campion (*Silene acaulis*); this has a woody rootstock that gives rise by repeated branching to a compact hemispherical cushion. At the top of the cushion are many very tiny green leaves, while below are the dark brown leaves of previous years. The plant has so many small rose-purple flowers that the cushion completely changes colour at flowering time. Another cushion plant is the mossy cherleria (*Cherleria sedoides*) which belongs to the same family, the Caryophyllaceae, as the moss campion. Its cushion is yellow green and its flowers are much less conspicuous than those of the moss campion being without petals and showing only tiny green sepals surrounding the stamens and styles. The flowers are regularly self-pollinated so the plant sets seed freely. An interesting feature of the plant's distribution is that it is one of the few alpines that does not occur in the Arctic. One really splendid cushion plant is the purple saxifrage (*Saxifraga oppositifolia*) which has opposite leaves and large red purple bell-shaped flowers. It flowers early, amid the melting snows of April and May, and it can be found at lower altitudes than many other arctic alpines. There are also other species of saxifrage that form rosettes that are only found on the Scottish mountain tops and they include some of the rarest British plants.

Another rare but very beautiful plant is the alpine forget-me-not (*Myosotis alpestris*); it is easily recognised. We have too an alpine gentian (*Gentiana nivalis*) an exquisite little plant that is one of the few annuals of the mountain tops and easily known by its intense blue flower borne on a stem a little over an inch in height.

On the more granitic mountains with their acid soils, shrubs of the heath family are to be expected and above the tree line heather moor and bearberry (*Arctostaphylos uva-ursi*) are often abundant. A carpet plant of the summit slopes is the trailing azalea (*Loiseleuria procumbens*) which often forms large colonies and bears numerous pink flowers. Alpine bearberry (*Arctous alpinus*) is deciduous and its leaves turn a bright red before they fall in autumn; the flowers are white borne in clusters of two or three, and they are succeeded by round black shining drupes that are eaten by birds that disperse the seeds.

Below the summits and in more favoured situations there may be other species, which may be thought of as plants of mountain pastures though the area of mountain pasture may well be very tiny. There are plants like the mountain everlasting (*Antennaria dioica*) and some hawkweeds (*Hieracium alpinum* and related spp.) and also the rock cresses (*Arabis petraea* and *Arabis alpina*) though the latter prefer the drier and more rocky situations. In more grassy situations one may come across the yellow mountain violet (*Viola lutea*) and the viviparous persicaria (*Polygonum viviparum*) which is like the persicaria of the lowlands but which bears purple bulbils in the lower parts of its inflorescence. The ladies mantles (*Alchemilla* spp.) may also be seen in these situations with their palmate leaves and spikes of small greenish flowers. Teesdale is particularly rich in these plants.

In the wet pastures and on damp rocks one can frequently find the yellow mountain saxifrage (*Saxifraga aizoides*) which often grows in quite large patches. It can be recognised by the deep yellow spot at the base of each petal and the bright red anthers. Another saxifrage to be found in wet places is the starry saxifrage (*Saxifraga stellaris*) which has a small rosette of oval-shaped leaves from which it sends up a panicle of small white flowers. Two yellow spots are found near the base of each petal.

Besides the arctic alpines there are many lowland plants that can be found high up on the mountains. Heather and bilberry are common enough in the lowlands and ascend to 3500 ft or even higher. The globe flower (*Trollius europaeus*) and the wood cranesbill (*Geranium sylvaticum*) together with angelica (*Angelica sylvestris*) can be found in damp meadow-like places on mountain slopes. Red campion (*Silene dioica*), Harebell (*Campanula rotundifolia*) and ox-eye daisy (*Chrysanthemum leucanthemum*) are to be seen on many mountain ledges. Man is responsible for carrying a weed flora with him wherever he goes. Thus the common stinging nettle (*Urtica dioica*) grows well enough on many mountain slopes and in the Alps it can sometimes be seen growing outside some of the high mountain refuges.

Of special interest are a number of mountain plants that also occur by the seaside. One may well be surprised to find thrift (*Armeria maritima*), sea plantain (*Plantago maritima*), scurvy grass (*Cochlearia officinalis*), sea campion (*Silene maritima*), and roseroot (*Sedum rhodiola*) growing in mountain pastures. Thrift is known by its dense cushion-like tufts of very narrow leaves that bear numerous flower stalks that are crowned with globular heads of pink flowers; sea campion has many creeping stems and leaves with a greyish appearance. The many white flowers are borne on separate stalks, and each petal has a black spot at the base. Roseroot possesses thick underground stems, while above ground, the shoots bear numerous fleshy leaves and heads of bright yellow flowers. This species is related to the houseleeks and stonecrops, many of which are mountain plants. Sea plantain is very like the common plantains of our gardens, but the leaves are long, narrow and fleshy. Scurvy grasses belong to the Cruciferae and the flowering stems grow from a loose rosette of circular to wedge-shaped leaves. The flowers are white in a loose spike and the fruit is a swollen almost spherical type of pod. There are four British species which are very similar and plainly closely related. The common species, *Cochlearia officinalis*, is particularly variable and the maritime and sub-alpine populations have been distinguished as sub-species.

This seems the place to refer to the very special flora of Upper Teesdale. Upper Teesdale is a relatively inaccessible and mountainous region lying between the Lake district and Durham. The highest point of the area is Crossfell which is just under 3000 ft. The whole district is of great interest to botanists and there are a variety of different habitats both on the lower and the upper ground. Much of Upper Teesdale is covered with blanket bog with its characteristic vegetation, but in places it is remarkable for a unique type of rock called sugar limestone which supports a unique flora. Among the rare plants of this area are the hoary rockrose (*Helianthemum canum*), the Teesdale violet (*Viola rupestris*) and a milkwort (*Polygala amara*) besides many lime-loving species like the blue grass (*Sesleria coerulea*) the meadow oats (*Helicotrichon* spp.) and some sedges. The best-known and most lovely plant of Upper Teesdale is the spring gentian (*Gentiana verna*) but it is

FIG. 46. Spring gentian (*Gentiana verna*).

impossible to do justice to the beauty of the area and its plants in any words.

It is generally considered that the plants of the Teesdale assemblage are relics of the late glacial flora of 12,000 to 15,000 years ago – a flora that was widespread over the whole country during that period. Later, changes of climate took place and the country was colonised by forest trees, in the earliest periods by birch and pine and later by trees such as oak, elm and lime. The forest, as it were, drove out the late glacial plants to the few places where the trees could not grow like the tops of mountains, the sea coasts and very rocky places. Thus plants like the thrift and other maritime species mentioned earlier which were widespread in late glacial times were driven to the mountain tops for survival. Conversely some of the mountain plants like the mountain avens were forced to the limestone areas near the west coast of Ireland and the north-west of Scotland where they still survive. In the intervening period, some species like the scurvy grass have diverged into mountain and coastal forms.

Another factor that has restricted these plants is their need of suitable habitats and their need to avoid too great competition. Much of the upper parts of the mountains became covered with blanket bog in which these plants cannot grow while in other places the competition may have been too severe. So only on those exceptional soils like the sugar limestone of Teesdale, the limestone pavement of the Burren in County Clare, and the open habitats of the banks of rivers could these plants survive.

If this hypothesis is correct these assemblages of plants are unique and of outstanding interest to the plant researcher. It is not surprising that the proposal of the Tees Valley and Cleveland Water Board to make a reservoir at Cow Green aroused strong opposition among botanists. £40,000 was raised to fight the case, but despite all the effort, the battle was lost and the reservoir has since been built. Time alone will tell in what way it will affect the area, whether, for example, the changes in the microclimate it has brought about will affect the flora, or whether the flow of the Tees itself will be so altered that the river banks no longer support their special plants. Upper Teesdale is a national treasure that we should do our best to maintain.

The lesson to be learnt from this fight is that no part of our national heritage of wild life is safe. The wilds of Upper Teesdale seemed remote and safe enough as do the mountain tops of Scotland. Man has not yet found a use for mountain tops, but constant vigilance is necessary so that our mountain flora is left for those that come after us to enjoy and study.

CHAPTER 16

SAND DUNES

SAND dunes make one think of the seaside and it is where most of them are to be found but smaller areas of blown sand do occur inland as in the breck country of Norfolk and Suffolk, and there are many species of plant that are common to both inland and seashore dunes. At the seaside, some of the first plants to grow on the seashore are the sea rocket (*Cakile maritima*) with pink to purple flowers and fleshy rather lobed leaves and the prickly saltwort (*Salsola kali*) a much branched annual with small greenish flowers, and sea sandwort (*Honkenya peploides*) a creeping plant with very small fleshy ovate leaves and greenish white flowers. These plants can grow only if the sand is reasonably stable and so they occupy the relatively narrow zone between the upper tide limit and the region of the more stable dunes. Their position is hazardous for an exceptionally high tide or storm may sweep them all away, after which a fresh start has to be made when the area is once more sufficiently stable. The plants, too, must be able to tolerate the presence of salt both in the sand and in the air. Plants able to do this are known as halophytes.

Plants and other obstacles slow down the wind that passes by them so that it will drop any fine particles of sand it may be carrying along. In this way a little mound of sand accumulates round the obstacle and it may grow until it is level with the height of the obstacle. Plants like those mentioned will in this way form miniature dunes but since most of the foreshore plants are annuals, this will not go on for a long time. There are some perennials like the sea couch grass (*Agropyron junceiforme*) and the sea lyme grass (*Elymus arenarius*) which can grow through accumulating sand and keep pace with it so that a very small dune may be found on the foreshore.

The great sand dune former is marram grass (*Ammophila arenaria*) which has a unique power to spread through sand by virtue of its numerous underground stems; moreover, it is able to grow through and keep pace with the accumulating sand so that dunes of even thirty to sixty feet high can be formed. It is relatively intolerant of sea water so it does not grow on the foreshore within reach of the tides, but a little further inland. The dune stabilising power of marram grass has been

noted and it has been planted on many coast lines to aid the building of
sea walls. There have been laws in the past forbidding the removal of
marram and also forbidding trespassing on dunes for fear of damaging it.

Once the sand surface has been partly stabilised by the marram,
other plants can establish themselves and aid the process. There are
grasses like the sand fescue (*Fescue rubra* var. *arenaria*) and sedges like
sand sedge (*Carex arenaria*) both of which have long creeping rhizomes
and there are deep-rooting plants like the sea holly (*Eryngium maritimum*), the sea birdweed (*Calystegia soldanella*) and also the sea

FIG. 47. Two plants of sand dunes: *left,* sea bindweed (*Calystegia
soldanella*); *right,* sea spurge (*Euphorbia paralias*).

spurge (*Euphorbia paralias*) that grows to about two feet and is full of
the milky juice so characteristic of the spurge family.

As the dune surface becomes more stable, many other species arrive
including plants that are common elsewhere like the ragwort (*Senecio
jacobaea*) and rest harrow (*Ononis repens*). The latter plant with its
branching stems, and pink pea-like flowers is common on dunes and is
able to avoid burial fairly successfully. Plants like the ragwort that
spend the first winter as rosettes may get covered with sand, and this is
probably why they do not occur in the earliest stages of dune formation.

There are a number of small annuals that occur on sand dunes. They are mainly winter annuals for they germinate in autumn and flower the following spring thus avoiding the summer when the dunes can become very dry indeed. These plants are well worth looking for and include the tiny scorpion grasses or forget-me-nots (*Myosotis* spp.), the chickweed (*Stellaria pallida*), the sand cat's tail grass (*Phleum arenarium*) and many others. Other annuals that are common weeds of cultivation like the scarlet pimpernel (*Anagallis arvensis*) can also be seen on sand dunes. In passing, it should be realised that such weeds owe their abundance to man's activity in cultivating and disturbing ground; before this occurred many were quite rare and restricted to open habitats like sand dunes.

When the surface movement of the sand has become virtually nil, the plant carpet begins to be complete and the growth of the small annuals reaches its peak. In many parts of the British Isles a great part is played by mosses and lichens in stabilising dunes and the latter are responsible for the colour change from the earlier yellow sand dunes to the older grey fixed dunes. The number of species that are found increases, and amongst them may be found hound's tongue (*Cynoglossum vulgare*) with dull red-purple flowers and rough greyish leaves. The whole plant has a mouse-like smell that readily identifies it. The stork's bills, members of the Geraniaceae, with their pink or white flowers and long pointed fruits are well represented. Rare sea stork's bill (*Erodium maritimum*) has simple lobed leaves while all the others have pinnate leaves. Some shrubs may make an appearance; very characteristic is the sea buckthorn (*Hippophae rhamnoides*) with narrow leaves that are silvery underneath. Its flowers are inconspicuous but the fruits are a beautiful orange and make the whole plant particularly attractive in autumn. The common bramble (*Rubus fruticosus*) is often an abundant duneland plant, but less common is the burnet rose (*Rosa pimpinellifolia*) which is a very spiny prostrate shrub that is spread by suckers. The flowers are creamy white and the fruits are purplish black and globular.

Dunes are established in successive series and between them there develop the successive 'lows' or 'slacks' that are sheltered and that retain more moisture than the dunes proper. The lows, too, benefit by drainage from the nearby higher dunes and not surprisingly support a characteristic flora. Plants like the creeping willow (*Salix repens*) and some rushes (*Juncus* spp.) occur freely in the early stages. Occasionally the slacks near the sea become flooded by a very high tide and then a brackish low is produced where plants like the sea lavenders (*Limonium*

spp.) and the sea milkwort (*Glaux maritima*) may be found together
with the sea heath (*Frankenia laevis*) which is not a member of the
heath family though it superficially resembles one. Many of the dune
slack species are the plants of damp places anywhere, for instance, the
marsh pennywort (*Hydrocotyle vulgaris*) and the lesser spearwort

FIG. 48. Two plants of the slacks of a sand
dune: *above,* sea lavender (*Limonium
vulgare*); *below,* marsh pennywort (*Hydrocotyle vulgaris*).

(*Ranunculus flammula*). Where there is standing water colonies of the
common reed (*Phragmites communis*), the yellow flag (*Iris pseudacorus*)
and the burr reeds (*Sparganium* spp.) may be seen. The richest dune
slacks contain some wonderful plants; for example the dunes of parts
of Lancashire where grass of parnassus (*Parnassia palustris*) and the
round-leaved wintergreen (*Pyrola rotundifolia*) flourish together with
several species of orchid.

Dune sand nearly always contains an amount of calcium carbonate
that has been derived from the remains of the shells of sea animals.
Where the dune sand is in part derived from calcareous rocks, the
amount can be as much as 70%; in many young dunes it is of the order
of 10–15% but occasionally it is very low as in the dunes of Studland

in Dorset. The result of this is seen in the abundance of many chalk-loving species in duneland habitats. Fairy flax (*Linum catharticum*), the perfoliate yellow-wort (*Blackstonia perfoliata*) and even some of the chalk orchids, like the scented (*Gymnadenia conopsea*) and the pyramid (*Orchis pyramidalis*), may be found in such situations. There are many others, the pink-flowered centauries (*Erythraea* spp.), the carline thistle (*Carlina vulgaris*), the purple fleabanè (*Erigeron acris*), some of the eye-brights (*Euphrasia* spp.), all of which can be considered as fairly typical chalk plants.

As, however, the dunes get older, the calcium carbonate gets washed out of the dune soils and a slow change in the soil reaction takes place. Professor Sir Edward Salisbury has been able to date some of the dune systems of England from old maps and relate these dates to the calcium

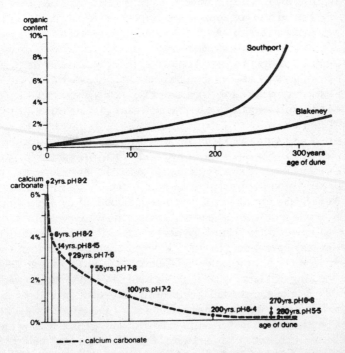

FIG. 49. Changes in the soil of sand dunes with age. The upper graph shows the increase in organic material with age; the lower shows the decrease in the content of calcium carbonate and the changes in soil reaction. After Salisbury.

carbonate and acidity of the soils. He has been able to show that the leaching of the calcium carbonate has been quicker on the west than on the east coast and that this is related to the greater rainfall on the west side of England. Thus the percentage of calcium carbonate in the dune soils of Blakeney in Norfolk fell from .4% to .1% in approximately 200 years; the fall in the Southport dunes was from 6% to .5% in about the same period of time. With these changes goes a corresponding increase in the acidity, the pH changing from 8.2 in young Southport dunes to 6.8 in the oldest.

Not surprisingly in the oldest dunes or where the dunes have contained very little calcium carbonate in the first place, plants that are favoured by acidic conditions are likely to be seen. On dune heaths like those of Studland in Dorset, plants like heather, the cross leaved and purple heaths may be found in abundance.

Extensive dune systems round our coasts are often taken over by man and turned into golf courses or afforested with conifers. But if they are not interfered with and left to themselves a woodland may be established. Privet (*Ligustrum vulgare*), blackthorn (*Prunus spinosa*), hawthorn (*Crataegus monogyna*) and hazel (*Corylus avellana*) are to be seen growing with oak on the dunes at Braunton in Devon and in time there may well be oak woodland. Elsewhere subspontaneous pine (*Pinus sylvestris*) colonises the heather areas and so again woodland may be developed.

One of the most successful dune afforestation schemes has been planting of the Culbin sands in Scotland which was started in 1922. The Culbin sands comprises an area some six miles long by two miles wide lying on the southern shore of the Moray Firth between the Moray–Nairn county boundary and the mouth of the River Findhorn. At one time, the whole was a fertile estate with many farms, but in 1694 it was overwhelmed by sand storms of great violence. The disaster may have taken place over a period of years rather than on a single occasion, but thereafter the Culbin Sands were a threat to the neighbouring rich farming land, for the sand could move rapidly under the influence of the prevailing west wind and cover fertile land in a night. The sands were originally stabilised by the growth of marram grass but the crofters gathered this in large quantities for thatching their crofts, and were apparently unaware of its value as a sand binder. So they helped to bring about the disaster that destroyed their livelihood. In the year after the disaster, the Parliament of Scotland passed a law forbidding the gathering of marram for thatching – a law that has never been repealed.

Some landowners made efforts to reclaim their land and achieved a limited success by planting trees but real progress was not made until the Forestry Commission started work in the 1920s. The major problem was how to stabilise the sand surface, and though at first marram grass was used, a quicker method known as thatching was used. This consisted of laying a covering of brushwood obtained from older plantations by cutting the lower branches of the trees over the surface of the sand. On the most exposed places it was pegged down with wire and wooden stakes, and in addition, birch, broom and smaller shrubs were used as they became available from the plantations. This material not only stopped the sand from moving, but as it decayed it became incorporated in the sand and formed humus.

Scots pine (*Pinus sylvestris*), Corsican pine (*Pinus nigra* var. *laricio*) and lodgepole pine (*Pinus contorta*) have been planted directly in the 'thatched' areas and have proved the best trees to grow. All three have the needles in pairs, and in the Scots pine they are about two to three inches in length. Corsican pine has longer needles (four to six inches) and they are a darker green than those of the Scots pine. Lodgepole pine has short needles (one and a half to three inches) but the cones have sharp projections on the scales that are lacking in the other two species. Corsican pine is better suited to the sandy soil and the salty air than any of the other trees; Scots pine and lodgepole are planted in frost hollows or on shingle where the Corsican pine does not grow so well.

Afforestation at Culbin must be counted as one of the success stories of the Forestry Commission. Any possibility of a repetition of previous disasters to good farmland has been prevented, useful timber is being grown, and efforts made to provide recreation for the passing visitor as well as for the naturalist.

CHAPTER 17

SHINGLE BEACHES

MUCH of our coastline consists of shingle beaches and they are not very hospitable places for plants to grow. Shingle consists of rock material that has been thrown up by the waves again and again so that it has become broken down into the smooth rounded pebbles and stones that are so familiar. Shingle beaches are formed in various ways. When shingle is carried on to a low-lying shore a fringing beach is formed and this kind of beach can be seen very frequently on the south coast of England. The shingle forms a ridge parallel to the shore, and most of the vegetation there is found on the landward side of the crest. A second kind of shingle beach is the shingle spit which is produced where there is an abrupt change in the direction of the coastline, while the sea currents continue moving in their original direction dropping shingle as they go. Spits of this kind may be seen at Blakeney in Norfolk, Calshot in Hampshire and Spurn head in Yorkshire. Sometimes the tip of the spit is deflected landwards forming a hook, and a series of hooks

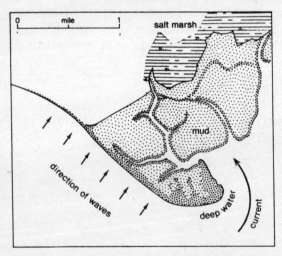

FIG. 50. The formation of a typical shingle spit.

may be formed one after the other. Salt marsh may form between the laterals, but in any case the shingle becomes more stable since it is protected by the outermost hook, and it therefore supports a more varied vegetation.

There are also shingle bars, that is shingle beaches running parallel to the shore, and joined to the land at each end, often enclosing a back-water. These are formed when a shingle spit stretches out across a bay and joins the land at the other side. One of the most striking examples is Chesil beach in Dorset which is over ten miles long and enclosing a long narrow lagoon known as the 'fleet'.

There is also a type of beach called an apposition beach when the shingle piles up in front of the waves and which is then driven by on-shore gales to form a bank parallel to the sea and out of reach of the tide. If this is repeated, a number of roughly parallel ridges is produced and a large area of stable shingle results as can be seen at Dungeness in Kent.

Whether plants can grow on the shingle depends on its degree of stability. Since there are many forces tending to keep the shingle moving which may vary from place to place, the stability is also very variable, and this accounts for much of the patchiness of the vegetation. There is very little humus in a shingle beach and what there is comes from the breakdown of plants and animals washed up by the tides; it is from this that the plants obtain the mineral salts they require for growth. Curiously enough shingle beaches are not short of fresh water, and shingle plants do not suffer from drought, for there are supplies of fresh water below the surface on which plants can draw. The most likely explanation of the source of this water is internal dew formation. On hot days the shingle warms up and so some air is driven out from between the stones and this air is replaced by cool air from above and also partly by air from further down in the shingle. When the stones cool at night, water is condensed from the relatively moisture-laden air around them. Although the level of water in shingle fluctuates, there is very little mixing between the salt water below and the fresh water above.

There are comparatively few plants that grow on shingle, which is not so surprising for so barren a place. Most of them are found on the more sheltered side of the shingle ridge, and there are usually a few on the ridge itself. One important plant is the sea campion (*Silene maritima*) already mentioned; it has deep roots and its large cushion-like growths help to stabilise the shingle. The yellow-horned poppy (*Glaucium flavum*), with its long curved seed pods, glaucous foliage and yellow

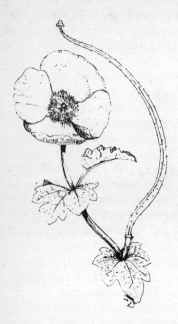

FIG. 51. Yellow horned poppy
(*Glaucium flavum*).

flowers is very characteristic as is the seaside form of the curled dock
(*Rumex crispus*). The woody nightshade (*Solanum dulcamara*) often
forms prostrate forms on shingle with fleshy unlobed leaves very
different from the more familiar types of the hedgerow and woodland.
The sea pea (*Lathyrus japonicus*) a rare plant of the south and east
shingle beaches has purple flowers that are always a delight to see.
Another shingle plant is the sea kale (*Crambe maritima*) which is a
large cabbage-like plant with white flowers and round globular fruits;
shoots from the rootstock of this plant grow through shingle quite
readily.

   One of the most remarkable plants of the shingle banks of the south
and east of England is the shrubby sea blite (*Suaeda fruticosa*) which
grows to a height of two to three feet and often forms a line along the
level of the shingle corresponding to that reached by the highest
spring tides. The seeds are dispersed by salt water and carried to the
highest tide mark where they germinate. The bushes, in a way similar
to marram grass, withstand a certain amount of burial by shingle and
can grow through it to make fresh growth a little higher up the slope.
Sea blite does not stand up to competition when growing on sand or
mud and so it tends to remain confined to the shingle.

Herb Robert (*Geranium robertianum*) is a common plant that establishes itself on shingle forming a basal rosette in the first season and flowering during the second year. A member of the Geraniaceae, it may be distinguished by the pink flowers, and palmate leaves with five leaflets and strong unpleasant smell. A much rarer plant is a very similar relative, *Geranium purpureum*, which is rather more hairy with duller flowers and smaller petals. The two species should be compared in order to appreciate the slight differences between them. *G. purpureum* is occasionally found on some of the south and west coasts of England.

Wild beet (*Beta maritima*) is also a shingle plant and there are one or two species of orache (*Atriplex* spp.) to be found on shingle; the latter are somewhat unattractive plants that lie close to the surface with toothed or lobed leaves and with the tiny flowers enclosed within two persistent green scales (bracteoles). The sea couch grasses (*Agropyron* spp.) are also to be seen growing through shingle; *A. junceiforme*) is more frequent on sand while *A. pungens* is more characteristic of the upper parts of salt marshes. From these places, both can invade nearby shingle.

Beaches are usually very exposed places but shelter is often provided on many by beach huts and similar seaside buildings; by these places rather taller growing plants like alexanders (*Smyrnium olusatrum*) may be found.

Plant succession on shingle does occur to the extent of forming a close surface cover. Very often it is the plants from the nearby marshes and dunes that colonise the older and more stable shingle such as the sea holly (*Eryngium maritimum*), the sea sandwort (*Honkenya peploides*), the scentless mayweed (*Tripleurospermum maritimum*), and the sea bindweed (*Calystegia soldanella*) though there are many inland plants that can grow successfully in these situations. Some grasses, the groundsels, the chickweeds and plantains are perhaps among the commonest of these.

Shrub colonisation by blackberry (*Rubus fruticosus*), blackthorn (*Prunus spinosa*), elder (*Sambucus nigra*), gorse (*Ulex europaeus*) and willows (*Salix* spp.) may also occur. On Dungeness near the Sussex-Kent border there is an extensive area of shingle vegetation where a particular dwarf variety of the broom (*Sarothamnus scoparius*) is to be seen which retains its characters in cultivation. Most interesting of all is the occurrence of holly (*Ilex aquifolium*) which, together with some other species, forms thickets that have been known there for a very long time. The part of Dungeness where the hollies grow is known as Holmstone, and the earliest certain date of its mention is by J. Leland in 1539, but there are several earlier references to woods on Dungeness, including one from a Saxon charter of 741, so they may be as old as 1230 years. Holly itself is long lived, and has maintained itself by natural regeneration until recent years; as shingle vegetation that of Holmstone is unique.

CHAPTER 18

SALT MARSHES

SALT marshes occur at the mouths of rivers or within the shelter of shingle bars or headlands. It is important to realise that without protection from the force of the sea the mud would be swept away. The fine particles of mud and sand that make up the marsh are carried to the river mouths where the water meets the incoming tides. The tide may rush in at first but at high water there is a short period when the water is very still and this is the time when the water drops its silt and the salt marsh begins to form. There are often seaweeds and other small plants that obstruct the flow and aid the tipping of the silt, and such obstacles also help to direct the water as it drains off the mud as the tide falls. A pattern of drainage channels is formed leaving the muddy areas exposed for colonisation by plants, which in turn trap more and more silt. A feature of a salt marsh is the presence of depressions filled with salt water and known as salt pans. They may be caused by uneven colonisation of the area in the first place, or by the stopping up of a drainage channel. The latter may happen by the plants growing completely over it, or by the collapse of the walls of the channel. As water evaporates from these pans a very strong solution of salt is formed, and when these dry up crystals of salt may be seen in the mud. Such areas are largely free of vegetation for the high salt content and the poor aeration greatly hinder the germination of seeds.

All salt marshes are regularly submerged by the tides, but the highest parts are only submerged by the spring and autumn tides, whereas the lowest muds are submerged twice a day. Germinating seeds will not become established if frequently submerged for they will be rolled over and over by the water. A minimum time without disturbance is required for the establishment and rooting of a seedling to the point where it can withstand some immersion without being washed away. In the case of glasswort (*Salicornia* spp.) one of the first colonists, this is about three days.

The plants of salt marshes must be able to tolerate salt and to grow in salt water. Plants able to do this are called halophytes, and some have a partial resemblance to the succulents of dry soils and desert habitats. This resemblance is quite superficial, for halophytes wilt readily if

removed from water and they do not store it in the way that some
desert plants do. Moreover, it has been shown that they remove water
without difficulty from the marshes in which they grow and lose it to
the atmosphere at a rate comparable to other plants. Salt is clearly the
most important factor that affects the plants of a salt marsh, and the
periods of immersion another, but it is possible for the plants of a marsh
to suffer from a physical drought at certain times. This can occur during
the summer, which is the period of the least rise and fall of the tide
and when the upper reaches of the marsh are beyond the reach of the
tide. If there is no rain at such a time drought may well occur. Other
factors that will affect the growth of the plants will be aeration of the
soil which can be very low as well as the nature of the soil itself. Some
plants for example, like the sea manna grass (*Puccinellia maritima*), are
favoured by a percentage of sand in the mud.

The first plants to grow on the surface of the bare mud are small
green seaweeds that are members of the green algae, and they cover the
mud with what looks like a green slime. They are not particularly
beautiful to look at with the naked eye, but viewed with the aid of a
microscope their true beauties are revealed and the colour, structure
and symmetry of the cells fascinate people when they see them for the
first time.

Sometimes eel grass or grass-wrack (*Zostera* spp.) is to be seen growing on the mud and this plant often extends to well below the level of
the low tide mark. Eel grass is a plant well beloved by ornithologists

FIG. 52. Glasswort (*Salicornia europaea*).

since it is one of the principal foods of geese and very large flocks are often seen feeding on the eel grass 'meadows'. It is a plant with long grass-like leaves that lie flat on the mud surface when exposed; botanically speaking, it is not a grass, but a highly specialised flowering plant that has adapted itself to survival in salt water, a thing which few flowering plants have managed to do. In addition, it has achieved a flowering process that takes place under water, the pollen itself being conveyed by water to the stigmas of the female flowers.

The first erect colonists of the mud of the salt marsh are the glass-worts or samphires (*Salicornia* spp.); these are green succulent and easily recognised plants though the exact delimitation of the individual species is difficult. Their small leaves are borne in pairs and are so fused along their margins that the axis has the appearance of being made of segments. The flowers are very small and partly enclosed by these segments. The species of *Salicornia* that colonise the mud are all annuals, and their stems and roots slow down the movement of water past them and so encourage the deposition of mud and the building of the marsh.

On some marshes, particularly those of the south cost of England, the mud is extensively colonised by cord grass (*Spartina townsendii*) which is a particularly effective stabiliser of mud even succeeding on mud too mobile for glasswort to colonise. The most interesting fact about this grass is that it was unknown until 1870 when it was first found in Southampton water and from whence it has spread with great rapidity along the coasts of Great Britain and it has also been planted in many parts of the world as a mud binder. The origin of this plant is very interesting, for it is the product of hybridisation between *Spartina stricta*, which is an uncommon native grass of mature salt marshes, and a North American cord grass (*Spartina alterniflora*), which was first noticed by Dr Bromfield in 1829 in Southampton water. Presumably this latter plant was brought from N. America by a ship and thrown overboard to become established in the Solent. *Spartina townsendii* is therefore a new species that has become established in the last century and its origin illustrates the part played by chance in distribution, for had not *Spartina alterniflora* crossed the Atlantic, *Spartina townsendii* would not have come into being.

As the mud is stabilised, many other species establish themselves, including other members of the Chenopodiaceae like the annual sea blite (*Suaeda maritima*) and sea purslane (*Halimione portulacoides*). The Chenopodiaceae are particularly tolerant of salt for members of this family are always prominent near sea coasts, salt mines, around oases and in salt deserts. Other species growing in the mud at this stage include the sea spurrey (*Spergularia maritima*) rather like a very hand-some and fleshy chickweed, the sea plantain (*Plantago maritima*), the sea arrow grass (*Triglochin maritimum*) that may be recognised by its narrow leaves and floral parts in threes, sea aster or michaelmas daisy (*Aster tripolium*) (not the garden species) which is a handsome plant growing two feet or so in height with fleshy narrow leaves. Usually it is easily recognised as a michaelmas daisy by its purple ray florets and yellow centres, but a variety does exist that lacks the purple ray florets and which is somewhat confusing to those who have not seen it before.

Often growing with the sea aster is the sea lavender (*Limonium vulgare*), and where it is well established it can turn the colour of a marsh to a lovely mauve when it is in flower. The common sea lavender has comparatively large leaves and the panicle of flowers is borne on a stem that is not usually branched below the middle. There are other species of *Limonium* that are less common and that usually grow higher up the salt marsh and on maritime cliffs and rocks; mostly they branch from near the base and have a rosette of basal leaves smaller than those of the common species. Above the sea lavender and sometimes growing with it there is often a zone of thrift (*Armeria maritima*). This plant has very deep roots by which it reaches a water supply that does not fluctuate either in quantity or concentration of salt as that near the surface. In these circumstances the rosettes have many long narrow leaves, but in drier ground the rosettes are much smaller in diameter and closely appressed to the ground. Thrift together with related species from abroad is frequently grown in gardens; so are species of *Statice*, near relations of the sea lavender, which are attractive by virtue of their persistent and brightly coloured calyces.

Sea purslane is a mealy shrub growing to three to four feet with a short creeping rootstock by which it can spread slowly. The leaves are oval and the plant bears several short yellow-brown flower spikes. It is found particularly where there is some sand mixed with the mud and where the marsh is well drained. It is seen along the sides of the channels that drain out of marshes, and often the course of such channels across the marsh can be traced by the silvery grey of this plant.

At the highest zone of a salt marsh two rushes are often seen. One is *Juncus gerardi*, which is a short rush usually less than a foot, with dark brown fruit, whereas the sea rush (*Juncus maritimus*) is taller with bright brown-coloured fruit. Other plants to be found in this zone include the sea wormwood (*Artemisia maritima*) with white downy leaves, the scurvy grasses (*Cochlearia* spp.) and grasses like the red fescue (*Festuca rubra*) and the sea couch grasses.

At the higher level in many salt marshes a gradual transition to freshwater marsh may be seen, but the existence of this does depend very much on local conditions. Plants that occupy such a transitional zone include the sea club rush (*Scirpus maritimus*) and also the common reed (*Phragmites communis*) which is tolerant of a small amount of salt in the water. A slender umbellifer (*Oenanthe lachenalii*), with white flower heads and finely divided leaves, occurs occasionally in such places and is worth looking for. Another very attractive plant of these marshes is the marsh mallow (*Althaea officinalis*) with pale pink flowers about two inches or more across, and with leaves that in earlier stages are folded like a fan.

The ultimate fate of a salt marsh at this late stage is usually enclosure and conversion to pasture, but there is no reason to doubt that the

Algal communities
    also eel grass (*Zostera* spp.)

Low marsh with
    glass wort (*Salicornia* spp.)
    cord grass (*Spartina townsendii*)
    sea manna grass (*Puccinellia maritima*)

High marsh with
    sea aster (*Aster tripolium*)
    thrift (*Armeria maritima*)
    sea lavender (*Limonium vulgare*)

Highest marsh
    sea rush (*Juncus maritimus*)

Reed flora (*Phragmites communis*)

Carr and woodland   salt pasture
                                     (if reclaimed)

FIG. 53. An outline scheme for a salt marsh succession.

freshwater marshy type of vegetation should not develop into some kind of fen carr and woodland (Chapter 11). Occasionally shrub colonisation by willows, bramble and the like can be seen at the extreme edge of such a freshwater marsh, and there are a few places round the Solway Firth where there is more convincing evidence. The transition from salt marsh to freshwater marsh is very well seen in some places on the Atlantic coast of North America, but the complete succession to mature forest cannot be seen anywhere. The treatment of the salt marsh in this chapter has been mainly from a successional point of view, so we include an outline scheme for the stages in such a succession on p. 135.

CHAPTER 19

THE RANGE OF BRITISH PLANTS

So far this survey has been concerned with plants growing in more or less definite habitats like woods, heaths, commons, or saltmarshes and the distribution of plant species as a whole in Great Britain has not been considered. Distribution is important for a little study will soon show that plant species show many different patterns of distribution even within such a small country as the British Isles. Some like the nettle-leaved bellflower (*Campanula trachelium*) are found only in or near woods in the south, while a close relative, the great bellflower (*Campanula latifolia*) occurs mainly in woods in the north of the country; similarly there are species like the oyster plant (*Mertensia maritima*) that are confined to sea shores in the north and a species of sea lavender (*Limonium bellidifolium*) that is confined to the east coast in England. Examples can be multiplied and their mention serves to indicate that all species have limits to their range and occupy definite areas of distribution. These areas may be very large for some plant species like the common reed (*Phragmites communis*) that are found almost all over the world or they may be very small, the species being confined to one small site a few small metres in area. The distribution of a species may be continuous when it is found within one whole area, but more often it is discontinuous occurring in several smaller areas with larger gaps between. There are many species e.g. the purple saxifrage (*Saxifraga oppositifolia*) that are to be seen on the mountains of Great Britain and which also occur on the alps of Europe and which do not occur in between. These are the so-called arctic-alpine group of the British flora. Similarly there is a group which occurs on the arctic mountains but not in the alps which is known as the arctic group.

In fact, all the species in the British flora have been placed in groups depending on their distribution outside Britain. There are those species which are centred round the Mediterranean and which just reach in England, and there are others with a marked oceanic and western type of distribution. In all, some sixteen groups of geographically related species have been prepared. It would be tedious to list them here, but the groups vary from a 'wide' element including species that are found in Europe, Asia and N. America and even further afield to a small group of endemics i.e. species known only from the British Isles. This latter group consists of plant species usually closely related to species belonging to one or other of the groups.

Within its range a species occupies only certain habitats. It may be restricted, say, to calcareous woodland like the green hellebore (*Helleborus viridis*) is in Great Britain or to alkaline fen rivers like the water soldier (*Stratiotes aloides*) of East Anglia. Alternatively the species may occur in a wider variety of habitats, like the common plantains (*Plantago* spp.) that are to be found on roadsides, waste ground, downland, commons and elsewhere. A distinction can therefore be made between the factors that affect the overall distribution of a species, and the factors that determine the kind of habitat it prefers within this area of distribution.

The flora of an area is often thought of as being rather static since plants are rooted to the ground and are not free to move about like animals. This is partly a false picture, for plants are in active competition with one another all the time and have numerous well-known devices by which they can spread from one place to another and thus invade fresh territories. In the British Isles some species are spreading, others are more or less stationary while many are declining. There have been several spectacular increases in the British flora within living memory. The Oxford ragwort (*Senecio squalidus*) was so named because of its occurrence as a rarity on some college walls at Oxford early in the thirties to which it had spread from the Oxford botanic garden. The species is really a native of Sicily where it grows on the slopes of Mount Etna. From Oxford the species gradually spread along the railway lines to London, where the damage done by air raids in the second world war provided an abundance of suitable sites for it to colonise. From these it has spread to the greater part of England and even Scotland as a plant of waste ground, roadsides and similar places. Many other examples can be quoted; the slender speedwell (*Veronica filiformis*) originally introduced from western Asia as a rock plant and first recorded as a garden escape in 1927 is now common in damp grassland in southern England and it too has spread well into Scotland. What is particularly interesting about this plant is that it very rarely sets seed in Great Britain so it must have been carried around by animals as very small pieces which possessed the ability to root and establish themselves very quickly. There are others that are

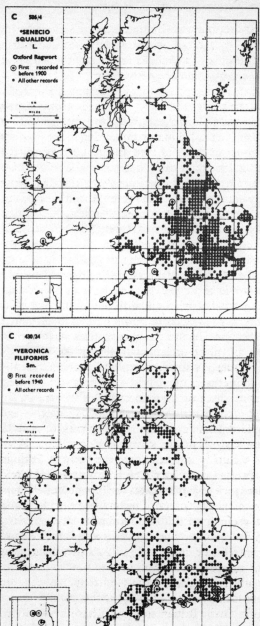

The distribution of Oxford ragwort, *Senecio squalidus* (above), and of the slender speedwell *Veronica filiformis* (below). These are examples of plants that have spread rapidly from localised sites. (Reproduced from *Atlas of the British Flora*, Ed. F. H. Perring and S. M. Walters, by kind permission of the Botanical Society of the British Isles.)

invading and spreading into the British Isles at the moment; in the first case they have been introduced by the agency of man, and they are spreading because man has so changed the face of the land that many new suitable habitats are now open to them.

On the other hand there are species that are declining and again the chief reason for this is that man has so changed or diminished their habitats that they can no longer survive. The great demand for water and for land drainage has led to the diminution in the number of bogs and marshy areas in England, and as a result many of the plants of wet places like the skull cap (*Scutellaria galericulata*) are less common than they used to be. Particularly striking too is the great decline in many cornfield weeds due to improved methods of seed cleaning, of cultivation and the use of synthetic herbicides. Corn cockle (*Agrostemma githago*) and cornflower (*Centaurea cyanus*) formerly common weeds are now rare and the common poppies are now rarely seen as scarlet sheets of colour in the fields.

The remainder and perhaps the majority of the flora may be said to be holding its own, and to be neither increasing or decreasing; provided woods are not felled, plants like the bluebell (*Endymion nonscriptus*), dog's mercury (*Mercurialis perennis*) and wood sanicle (*Sanicula europaea*) seem just about as frequent as they have been in the past. The reverse of this argument may also be broadly true, namely, that a species that is spreading rapidly has been recently introduced while a species that is more or less stationary or declining has been in the country a long time and is native. There are exceptions to this for man may change the environment so quickly and extensively that opportunities for spread are suddenly increased. Rosebay (*Chamaenerion angustifolium*) is a good example, for about fifty years ago it was a local plant found here and there throughout the country in rocky places and on open screes. Now it is widespread and very common on waste ground, heaths and commons, Perhaps in this case there are other contributory factors, for the plant may have undergone genetic change and so become more aggressive. Plants of course, do indeed show such changes and it well known that a plant species may not be morphologically and physiologically the same over all its range. Common broom (*Sarothamnus scoparius*) exists as a dwarf, rather prostrate, variety on the cliffs of south Cornwall and parts of the south coast. If this variety is cultivated it retains the prostrate habit and it may well be affected differently by the factors of the environment from that of the common inland form. The native Scot's pine (*Pinus sylvestris*) of the Caledonian forest is slightly different from the introduced *Pinus sylvestris* of southern England and these two forms may also react differently to environmental factors.

It will be seen that the understanding of the causes underlying the distribution of any one species of plant is not easy to come by but before this can be attempted, the area of distribution must be known. As far as the British Isles are concerned recording of wild plants has tended to be county by county, a practice which goes back to the days of John Ray who published a catalogue of the plants of Cambridgeshire in 1660. His example was followed by many others and accounts or lists for most counties have been prepared over the last three centuries. An approximate idea of the distribution of a species can be gained by seeing from which counties it has been recorded and then preparing a map on which the occupied counties are shaded. However counties differ greatly in size and a single record from a large county results in the shading in of a much larger area on the map than one record from a small county. To get over this difficulty, or at least to diminish it, H. C. Watson in 1873 divided up the larger counties so as to produce 112 smaller units more equal in size and these he called vice-counties. From these, maps of the distribution of British plants could be made by blocking out the vice-counties for which any species had been recorded. Such maps give no idea of the relative frequency of the plants within the vice-counties for, as said above, one or hundreds of records from a vice-county sufficed to block out the whole on the map. In fact the inadequacy of these maps stimulated the search for something better, and following on the production of dot maps for Holland and N.W. Europe, the Botanical Society of the British Isles in 1953 sponsored the preparation of a series of similar maps for Great Britain.

This project was aided by the introduction of the National Grid which was not available to the earlier workers. The grid is shown on all Ordnance Survey maps; they are ruled to show 10 km squares in darker lines and 1 km squares in finer lines. There are about 3500 10 km squares in Great Britain and between 1954 and 1960 the whole of the country was surveyed by botanists and the species present in each 10 km square recorded. To aid the recording, cards listing the species most likely to be encountered were prepared, the names being crossed through as the individual species were seen. From all these records, the atlas of the British flora was prepared and published by the Botanical Society in 1962. This was a great advance on anything that had gone before and it has stimulated

The distribution of yellow-wort, *Black-stonia perfoliata*, shown by vice-counties (above) and by 10 km squares (below). The latter, because of its greater precision, has become the standard method of mapping plant distributions. (Reproduced from *Atlas of the British Flora*, Ed. F. H. Perring and S. M. Walters, by kind permission of the Botanical Society of the British Isles.)

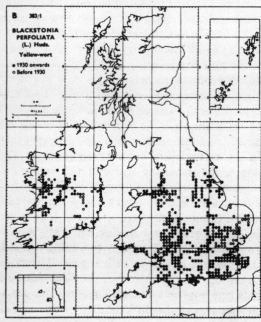

B   383/1

BLACKSTONIA
PERFOLIATA
(L.) Huds.

Yellow-wort

● 1930 onwards
○ Before 1930

more than 30 similar projects in other groups of organisms all now in various stages of completion.

Notwithstanding the success of the atlas, it was fully realised at the time that not all the 10 km squares were as completely surveyed as they might have been. Also the content of any 10 km grid square may well have changed since 1962 and so it is still possible for anyone to add to existing records by careful searching. You can record the presence of plants in your area and find whether or not you have a new record by consulting the atlas which is most likely available at your local library. It is hoped to make a fresh survey and to publish a new set of maps about 1985.

Survey on the basis of 10 km grid squares is more thorough than on a vice-county system, but all the same, the 10 km grid square is quite a large unit. So the atlas has been followed up by more intensive surveys based on smaller units. Usually these have been initiated for a county, but occasionally a well-defined vegetational area has been chosen like the Breckland of East Anglia. The unit chosen for survey has varied; the Warwickshire flora (1971) used 1 km squares; the Berkshire flora (1968) used 5 km squares. Others have used a 2 km square (four of the smallest squares on the 1:50,000 O.S. maps), colloquially called a 'tetrad'. The preparation of a county flora on this basis is a big task; Sussex, for example, has over 1000 'tetrads' to survey. Floras containing maps prepared on a 'tetrad' basis include Flora of Staffs (1972) and the Flora of Herts (1967). A number of others are in preparation and you may very well be able to help if you can get in touch with the county organisers through a local Natural History society or Naturalists' trust.

In all these studies the distinction between the actual area over which the plant is spread and the frequency of the plant within that area should be realised. Clearly the smaller the recording unit, the better the appreciation of the frequency of occurrence within the total area. Thus two species may occur, for example, in all the 10 km squares of Surrey but one may occur in a very few of the 'tetrads' and the other in nearly all. Some floras have attempted to indicate the frequency within the chosen unit by using various symbols. Berkshire has done this for the 5 km grid squares, indicating whether the plant is frequent occasional or rare in each square. The most complete information of this kind is given in the Warwickshire flora, the maps of which give not only an indication of the frequency of occurrence of each species, but also an indication of the type of habitat in which it is found.

CHAPTER 20

THE INTERPRETATION AND UNDERSTANDING OF PLANT DISTRIBUTION MAPS

THE preparation of distribution maps is interesting and valuable in itself, but the enquiring reader will wish to know the reasons for the many different patterns that are shown by the maps. These are by no means easy to find out, and they are hardly fully known for any single species for any particular interpretation needs to be checked by field observations, experiments and further research. There are pitfalls to avoid. For example, a map of a species that is actively spreading and extending its range will show a purely accidental distribution depending on when the original survey was made. This does not only apply to a plant species spreading rapidly; it may apply to a species that is spreading so slowly that in the comparatively short time that accurate distribution studies have been carried out, it has made imperceptible progress. Obviously plant species in this category are difficult to distinguish and a plant geographer may draw false conclusions.

One of the simpler patterns that shows up well is that of those plants confined to calcareous soils. One such plant is horseshoe vetch (*Hippocrepis comosa*), another is the rampion (*Phyteuma tenerum*) which is strictly limited to the southern half of England, and another which is rather more common is the stemless thistle (*Cirsium acaulon*). Clearly there are factors other than the calcium involved for it is necessary to account for the differing degrees of restriction to the calcareous habitats shown by these species. These are most likely climatic, for horseshoe vetch is nearly always found on south or south-west facing slopes where the season is long enough for the sun to warm the ground effectively and thus induce the flowering and ripening of fruit. Much the same is true for the stemless thistle which, although it occurs on slopes of different aspect in the south, becomes increasingly restricted to warm and dry slopes as it goes further north. In the south of England the thistle flowers from late June to September and sets fruit readily but at its northern limit in Derbyshire, it does not flower until August and it only sets fruit well in those years when August and September are warm and dry. Professor Pigott has been able to confirm these conclusions by ingenious experiments in which asbestos screens were placed from east to west across the centres of plants so that on the north side the flower heads and basal leaves were shaded from direct sunlight. These produced temperature differences between the flowers' heads on the north and south sides and there was a reduction in the proportion of viable fruits formed on the northern side of the screens. In addition some of the plants were sprayed with water each day and this increased the temperature differences to as much as 10°C under clear skies and the reduction in the number of viable fruits was then as much as 75 per cent. So in this example, there is clear evidence of the summer temperature working with the soil factors to account at least in part for the observed distribution. Before leaving calcicolous plants there are highland species that show the same restriction. A particular example is the rock speedwell (*Veronica fruticans*) which is restricted to calcareous rocks above 1500 ft in Scotland. It is an arctic-alpine plant and it occurs in Greenland, Scandinavia and the mountains of the Pyrenees, the Alps and the Caucasus. Such species cannot apparently tolerate the high summer temperatures of the lowlands.

In contrast to the distribution shown by the calcicoles, there are the calcifuges or lime-hating plants. The distribution of cross-leaved heath (*Erica tetralix*) shows it to be absent from the chalk and limestone areas, but it is otherwise throughout the British Isles and it extends further north on the continent into Scandinavia, further south into Spain and Portugal, and to the east into France, Germany and Poland. Therefore in England it is within its overall range and not limited by climate. Another less common calcifuge is the bog gentian (*Gentiana pneumonanthe*) which is confined to damp acid heathland; it is a lowland plant, not ascending to more than 800 ft in Britain and this would appear to be a climatic limitation. The discontinuous distribution shown by the map is accounted for by the relative lack of suitable habitats.

Turning more particularly to climatic factors it is obvious that they are so many and so complex in their interaction that it is unlikely that the range of any given plant can be explained entirely and simply by the action of a simple climatic factor. However, one thinks of winter cold as being a likely factor and the coldest places occur in east Scotland and north-east England. The twin flower (*Linnaea borealis*) is limited to these areas and seems to be intolerant of high winter temperatures. Maybe its seeds require exposure to a

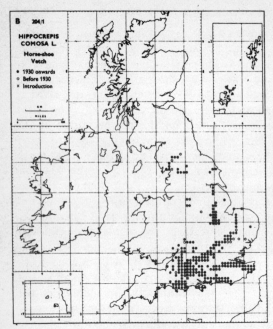

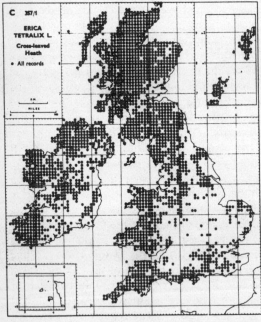

The distribution of horseshoe vetch, *Hippocrepis comosa* (above), and of cross-leaved heath, *Erica tetralix*, (below). The restriction of horseshoe vetch to calcareous soils, and the absence of cross-leaved heath from them, is easily seen. (Reproduced from *Atlas of the British Flora*, Ed. F. H. Perring and S. M. Walters, by kind permission of the Botanical Society of the British Isles.)

low winter temperature before they will germinate. There are other plant species which are the reverse of these and which cannot tolerate low winter temperatures. These are some of the plants of south-west England like the Dorset (*Erica ciliaris*) and Cornish (*Erica vagans*) heaths and the plants of south and west Ireland like the strawberry tree (*Arbutus unedo*) and St Dabeoc's heath (*Daboecia cantabrica*). Plants like the strawberry tree are frost sensitive and killed in severe winters, but there are plants that grow extensively in the winter and spring and are more or less dormant in the summer. Rusty back fern (*Ceterach officinarum*) and small-flowered buttercup (*Ranunculus parviflorus*) are perhaps examples, and if the winter is too cold they cannot grow well enough to survive.

The effect of high summer temperatures has already been mentioned in connection with the calcicolous species and just as there are species requiring high summer temperatures there are others that are intolerant of them. Some mainly northern species like the marsh cinquefoil (*Potentilla palustris*), the wood pea (*Lathyrus sylvestris*) fall into this latter category. Rainfall is obviously a climatic factor of importance and it may vary from 20–25 in in parts of eastern England to 150 in in parts of the Lake district. The low rainfall of the east combined with the naturally dry porous soils of the Breckland make it one of the driest parts of the country and not surprisingly there is a group of plants that are largely confined to the Breckland. These include the Spanish catchfly (*Silene otites*), two rare speedwells (*Veronica verna* and *V. praecox*) and the field wormwood (*Artemisia campestris*) and others. These areas resemble the steppes of parts of the continent where the plants also grow, and in consequence they have been referred to as the 'steppe' element in the British flora.

It is harder to point to species favoured by the greater rainfall and humidity of the western side of the British Isles because the differences in temperature and geology are also coincident with the differences in rainfall. However, it is difficult not to relate the greater frequency of ferns in the west as compared with the east to the greater humidity of those parts.

Of the many other examples that could be taken to illustrate the action of various factors of the environment we may take a plant like Jacob's ladder (*Polemonium coeruleum*) which is a rarity known from grassy slopes and ledges on limestone hills; it ranges from Derbyshire northwards to west Yorks and the Cheviot hills. In these places, it is restricted to northern slopes where hill mist is frequent, and when it grows at the foot of cliffs it is in places shaded at midday. Experiments have shown that the resultant reduction in temperature is important in reducing the water loss from the plant for it cannot withstand any summer drought. Now the pollen grains of this plant can be identified with certainty, and it is therefore of remarkable interest that fossil pollen grains have been recovered from several late Glacial and early post Glacial (Ice age) sites in England and Wales. It seems that the plant then grew in many places in the comparatively open vegetation that followed on the retreat of the ice from Britain. As the vegetation succeeded to forest, the plant became rarer and it was forced to a few sites where its essential requirements, moist fertile soil, freedom from competition and low summer temperatures could be met. The present-day distribution of this plant is to be regarded as consisting of a few relict areas remaining from the earlier wider distribution in the late Glacial period.

Jacob's ladder is an example of the importance of the historical factor in determining the present distribution of British plants. It is easy to neglect this factor but it is important both in the long and short term view. The comparative paucity of the British flora is due to the Ice ages, the last of which ended some 10,000 years ago. As the climate ameliorated, plants repopulated the British Isles in the wake of the retreating ice, making use of the land bridge between England and the continent. The separation of England from the continent some 7–8000 years ago produced a formidable barrier to the arrival of new species. Plants that have arrived since must have had a dispersal mechanism capable of crossing the Channel and the evidence is that they constitute but a small group, or, they must have been introduced by man. This latter is much the larger group and includes some plants species deliberately introduced by man and a much larger number of species introduced by accident. There are several plant species that (e.g. *Cirsium oleraceum*) flourish just the other side of the Channel in circumstances much the same as in south-east England but which do not occur naturally in England. The most likely explanation is that they did not reach England before the land bridge was broken and have since been unable to cross the Channel. It is worth mentioning that a few colonies of this plant which have been introduced into this country maintain themselves quite well.

The special Upper Teesdale flora has already been referred to as a relict flora dating from late Glacial time; here and there there are other relics of post glacial vegetation like the pine forests of parts of the Scottish highlands that date back to the warm, dry (boreal)

The distribution of twinflower, *Linnaea borealis* (above), and of Jacob's Ladder, *Polemonium coeruleum* (below). Both are examples of the influence of climatic factors, whether present-day or historical, in determining the distribution of a plant. (Reproduced from *Atlas of the British Flora*, Ed. F. H. Perring and S. M. Walters, by kind permission of the Botanical Society of the British Isles.)

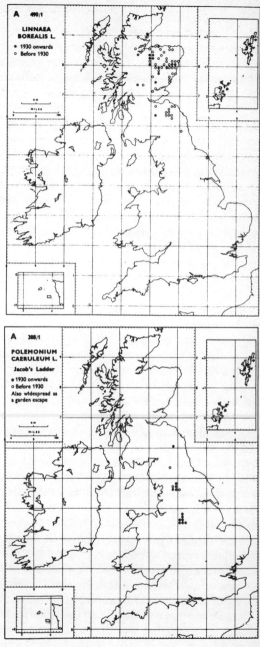

times that followed on the retreat of the ice. The history of other
woodlands is far more chequered and cannot be followed in any
detail in this book but there is good evidence for some woodlands
never having been clear felled by man, but having been, of course,
managed by man for his needs. Such woodlands have been termed
primary woodlands, and they are often rich in species, including
some with a very low rate of spread. The best of these woodlands
are well known and they are mostly protected by one organisation
or another. Other woodlands may have been planted relatively
recently, say 100–200 years ago, or they may have developed from
an abandoned plantation of trees or they may have just developed
from as a natural succession from abandoned land. Such woodlands have been called secondary. The field botanist wandering
from one piece of woodland to another adjacent to it may wonder
why one piece is rich in species while the other is poor. It could
be that one is more recent than the other, and many plant species,
owing to their slow rate of spread, have been as yet unable to colonise the newer woodland. In past times when the areas of woodland
in lowland Britain were much greater, the spreading of woodland
species would have been easier, but as woodland is cleared, isolation
of fragments increases and the distinctions between woodlands that
are related to their age tend to become perpetuated.

Post Glacial history also helps to explain some of the rather
curious plant distributions of the chalk of southern England. As
has been mentioned the major event following the retreat of the ice
was the development of a continuous forest cover over almost the
whole of the country. This was gradually cleared by Neolithic man
and his activity can roughly be dated to about 3000 BC. The cleared
areas were used by man for his agriculture and the chalk downlands
that we know today began to come into existence. This raises the
problem as to where the chalk grassland species came from (for
they are not woodland species) and it seems that they must have
been either introduced by man himself or they must have persisted
in a few specially favourable habitats all the time since their original
arrival following on the departure of the ice, and then subsequently
have spread into areas cleared by Neolithic man. The latter view is
perhaps the most likely, and some places have been suggested as
sites where species may have survived. One of these is the very steep
slopes above the river Mole at Box Hill in Surrey. Here the box
(*Buxus sempervirens*) together with other species may well have
survived right through the forest period from the time of their
original entry until today, and from where they may have spread to

the areas of cleared woodland nearby. Other likely sites that have been suggested are the chalk slopes of the Medway gap and cliffs along the parts of the south coast. In support of this view, there are a number of rare chalk loving species that are found in the neighbourhood of these sites which decrease in frequency as one goes away from them. Thus past history may provide part of the answer to the distribution of these species, but the above suggestions are but an unproven hypothesis at the moment.

Endemic British species – species that have originated in Britain – are few in number, as Britain has been an island only a short time. In islands of great age geologically, like the Canaries, endemics may include both last relics of species dying out and new youthful species just emerging. The flora of the Canaries amounts to 1700 species of which about 1000 are native and 470 are endemic. In Great Britain the percentage is much less, perhaps 2 to 3 per cent but in a flora as much studied as that of the British Isles and where the number of endemics is very small, the figure does depend very much on whether any particular variant is awarded specific or only subspecific status. The eyebrights (*Euphrasia* spp.) contain a number of endemics, two confined to the island of Rhum and one (*E. pseudo-kerneri*), which is a large flowered species, confined to calcareous downs in southern England. Another endemic of these downs is a small annual gentian (*Gentianella anglica*) which flowers as early in the season as May or June. Another interesting endemic is a rare fumitory (*Fumaria purpurea*) related to the ramping fumitory (*Fumaria capreolata*), but distinguished from it by its purple flowers and a number of other rather small characters and which is found entirely in artificial habitats like waste ground, hedgebanks, gardens and other cultivated ground. Presumably it could not have survived without the existence of these habitats.

Lastly there is a very local water dropwort (*Oenanthe fluviatilis*) a member of the Umbelliferae, which occurs in slow moving rivers and streams. It has finely divided leaves with emergent stems which bear typical umbels of white flowers. Like many water plants it can be abundant where it occurs, but otherwise it is quite uncommon. It is known from one or two places in Denmark and Germany, so it is not an endemic, but there is far more of it in the British Isles than on the continent. So maybe it could have originated in Britain and spread to the continent as a kind of botanical export; if so, and there is no proving it, it is the only plant species likely to have achieved this.

CONSERVATION

The need of land for housing, for factories, for airfields in addition to all other uses is increasing all the time and no more so than in southern England. As the number of people who live and work in this country increases so does the need of land for recreation, for games, for walking or just to supply the need to get away from the confinements of the daily round for a time. So, between all these needs, the places where wild plants and animals live are subject to great pressures and with the result that there has been a great decline in their numbers. This book has been concerned chiefly with wild plants and all the facts show that our native flora of 1500 species is diminishing and about 10% is in danger of extinction in the next few years unless more positive steps are taken to stop the decline.

It should not be thought that our rare plants are all far away and inaccessible in the wildest and most remote parts and that your own particular district contains no plants of any rarity or value. The reverse is the case; for example, Hampstead Heath in London still retains botanical interest and many an uncommon plant can still be found long after its surroundings have been completely changed. Many beauty spots visited by thousands of picnickers in summer sunshine are the homes of rare plants and animals, and some of the rare seaside plants must be near to many a holiday maker. It is therefore important to be awake all the time and in all places to the need for conserving our flora and fauna.

Numbers of people from the Government downwards are concerned with the state of our environment and some are specially concerned with those places of particular natural history and scientific interest. Almost certainly there will be a Naturalists' Trust in your county, and if you share the interests of this book, then make it your concern to support it by all the means in your power. The conservation of rare plants (or interesting associations of common ones) means establishing nature reserves which usually involves some money and often very large sums indeed; but it is not enough to purchase land for a nature reserve for it must then be managed so that conditions are kept as near perfect as possible for the organisms it is wished to conserve. This is where you may be able to help and this work is just as important as finding the money in the first case. It may mean cutting down scrub that is threatening the habitat of a chalk downland orchid or clearing out a pond that is silting up.

Rare plants only survive in very special habitats and hence their opportunities for spreading are very limited. Thus when a site is lost, it is usually lost for ever, for the species does not gain another in its place. The need for active conservation has never been greater. Support it then, in every way that you can.

A CODE OF CONDUCT

1 The first essential is to preserve the habitat. This people can easily and unwittingly damage.

2 When going to see a rare plant, avoid doing anything which would expose it to unwelcome attention, such as making an obvious path to it or trampling on the vegetation around it.

3 'Gardening' before taking photographs may also have this effect, and nearby plants can be crushed by kneeling photographers.

4 Remember that photographs can give clues to the localities of rare plants, quite apart from the information accompanying them.

5 Avoid telling people about the site of a plant you believe to be rare. Your local nature conservation trust, who will help safeguard it, should, however, be informed.

6 Respect requests from conservation bodies or land owners not to visit particular sites at certain times.

7 The uprooting of wild plants is to be strongly discouraged. Most local authorities have bye-laws against this, so it may be illegal.

8 If living plants are needed for cultivation, take seed or cuttings sparingly, and not from those that are rare.

9 Pick only flowers known to be common or plentiful in the locality.

10 No specimens should be taken from any nature reserve or National Trust property.

11 Plants should not be introduced into the countryside without the knowledge and agreement of your local nature conservation trust or natural history society.

BOOKS TO READ

For general reading, *Britain's Green Mantle* by A. G. Tansley (revised by M. C. F. Proctor; Allen and Unwin) is a beautifully illustrated book affording an excellent introduction to British vegetation. *Wild Flowers* by J. S. L. Gilmour and S. M. Walters (Collins) and *Companion to Flowers* by D. McClintock (Bell) are full of information on the naming, history, distribution and folklore of British plants. For collection and identification, *The Pocket Guide to Wild Flowers* by D. McClintock and R. S. R. Fitter (Collins) is one of the easiest guides to the identification of wild plants. *The Wild Flowers of Britain and Northern Europe* by R. S. R. Fitter, A. Fitter and M. Blamey (Collins) is beautifully illustrated and well planned. The standard flora of the British Isles is *Flora of the British Isles* by A. R. Clapham, T. G. Tutin and E. F. Warburg (Cambridge); this uses scientific and technical terms extensively.

R. S. R. Fitter's *Finding Wild Flowers* (Collins) is a guide to places to search for wild flowers.

For trees and shrubs A. Mitchell's *A Field Guide to the Trees of Britain and Northern Europe* (Collins) is comprehensive and popular. Grasses, sedges and rushes are covered by McClintock and Fitter (above) but there are also W. J. Stokoe's *The Observer's Handbook of Grasses, Sedges and Rushes* (Warne; elementary) and C. E. Hubbard's *Grasses* (Pelican) and C. Jermy and T. G. Tutin's *British Sedges* (Bot. Soc. Brit. Isles).

For particular habitats there are many books including several in the *New Naturalist* series. For woodlands, T. R. E. Southwood's *Life of the Wayside and Woodland* (Warne) is introductory, J. D. Ovington's *Woodlands* (E.U.P.), C. R. Stubbs's *The New Forest* (David & Charles), W. Condry's *Woodlands* (Collins) all treat the subject from different angles. Chalk downlands are well covered by J. E. Lousley's *Wild Flowers of Chalk and Limestone* (N.N.); for heaths and commons and waste ground there are Stamp and Hoskins's *The Common Lands of England and Wales* and *Weeds and Aliens* by E. J. Salisbury, both N.N.; for marsh and fen there is A. Arber's classic work on *Water Plants*, eminently readable and available as a reprint; E. A. Ellis on *The Broads* is also informative on water plants. For the moors and mountains there are John Raven and Max Walters's *Mountain Flowers* (N.N.), W. Condry, *The Snowdonia National Park* (N.N.), W. H. Pearsall *Mountains and Moorlands* (N.N.) For coastal vegetation *Flowers of the Coast* by I. Hepburn (N.N.) makes good reading and for the history of British vegetation there is W. Pennington *The History of British Vegetation* (E.U.P.).

INDEX

Plate references are in **bold** type